Topics in Intelligent Engineering and Informatics 4

Editorial Board

Editors-in-Chief

János Fodor
Imre J. Rudas

Editorial Advisory Board

Ildar Batyrshin (Mexico)
József Bokor (Hungary)
Bernard De Baets (Belgium)
Hamido Fujita (Japan)
Toshio Fukuda (Japan)
Fumio Harashima (Japan)
Kaoru Hirota (Japan)
Endre Pap (Serbia)
Bogdan M. Wilamowski (USA)

Review Board

P. Baranyi (Hungary)
U. Bodenhofer (Austria)
G. Fichtinger (Canada)
R. Fullér (Finland)
A. Galántai (Hungary)
L. Hluchý (Slovakia)
MO Jamshidi (USA)
J. Kelemen (Czech Republic)
D. Kocur (Slovakia)
P. Korondi (Hungary)
G. Kovács (Hungary)
L.T. Kóczy (Hungary)
L. Madarász (Slovakia)
CH.C. Nguyen (USA)
E. Petriu (Canada)
R.-E. Precup (Romania)
S. Preitl (Romania)
O. Prostean (Romania)
V. Puri (Italy)
GY. Sallai (Hungary)
J. Somló (Hungary)
M. Takács (Hungary)
J. Tar (Hungary)
L. Ungvari (Germany)
A.R. Várkonyi-Kóczy (Hungary)
P. Várlaki (Hungary)
L. Vokorokos (Slovakia)

For further volumes:
http://www.springer.com/series/10188

Aims and Scope

This book series is devoted to the publication of high-level books that contribute to topic areas related to intelligent engineering and informatics. This includes advanced textbooks, monographs, state-of-the-art research surveys, as well as edited volumes with coherently integrated and well-balanced contributions within the main subject. The main aim is to provide a unique forum to publish books on mathematical models and computing methods for complex engineering problems that require some aspects of intelligence that include learning, adaptability, improving efficiency, and management of uncertain and imprecise information.

Intelligent engineering systems try to replicate fundamental abilities of humans and nature in order to achieve sufficient progress in solving complex problems. In an ideal case multi-disciplinary applications of different modern engineering fields can result in synergistic effects. Informatics and computer modeling are the underlying tools that play a major role at any stages of developing intelligent systems. Soft computing, as a collection of techniques exploiting approximation and tolerance for imprecision and uncertainty in traditionally intractable problems, has become very effective and popular especially because of the synergy derived from its components. The integration of constituent technologies provides complementary methods that allow developing flexible computing tools and solving complex engineering problems in intelligent ways.

Jozef Kelemen, Jan Romportl, and Eva Zackova (Eds.)

Beyond Artificial Intelligence

Contemplations, Expectations, Applications

Springer

Editors
Prof. Jozef Kelemen
Institute of Computer Science
Silesian University
Opava
Czech Republic

Dr. Eva Zackova
New Technologies Research Centre
University of West Bohemia
Plzen
Czech Republic

Dr. Jan Romportl
New Technologies Research
Centre & Department of Cybernetics
University of West Bohemia
Plzen
Czech Republic

ISSN 2193-9411
e-ISSN 2193-942X
ISBN 978-3-642-43738-0
ISBN 978-3-642-34422-0 (eBook)
DOI 10.1007/978-3-642-34422-0
Springer Heidelberg New York Dordrecht London

© Springer-Verlag Berlin Heidelberg 2013
Softcover reprint of the hardcover 1st edition 2013
This work is subject to copyright. All rights are reserved by the Publisher, whether the whole or part of the material is concerned, specifically the rights of translation, reprinting, reuse of illustrations, recitation, broadcasting, reproduction on microfilms or in any other physical way, and transmission or information storage and retrieval, electronic adaptation, computer software, or by similar or dissimilar methodology now known or hereafter developed. Exempted from this legal reservation are brief excerpts in connection with reviews or scholarly analysis or material supplied specifically for the purpose of being entered and executed on a computer system, for exclusive use by the purchaser of the work. Duplication of this publication or parts thereof is permitted only under the provisions of the Copyright Law of the Publisher's location, in its current version, and permission for use must always be obtained from Springer. Permissions for use may be obtained through RightsLink at the Copyright Clearance Center. Violations are liable to prosecution under the respective Copyright Law.
The use of general descriptive names, registered names, trademarks, service marks, etc. in this publication does not imply, even in the absence of a specific statement, that such names are exempt from the relevant protective laws and regulations and therefore free for general use.
While the advice and information in this book are believed to be true and accurate at the date of publication, neither the authors nor the editors nor the publisher can accept any legal responsibility for any errors or omissions that may be made. The publisher makes no warranty, express or implied, with respect to the material contained herein.

Printed on acid-free paper

Springer is part of Springer Science+Business Media (www.springer.com)

Preface

The book that you are about to read consists of selected papers which were originally presented in December 2011 at the conference *Beyond AI: Interdisciplinary Aspects of Artificial Intelligence* in Pilsen (Plzeň), Czech Republic.

Organisers of conferences and editors of the resulting proceedings often tend to write prefaces emphasising importance and scientific excellence of their respective conferences. I am not going to write anything like this here because I know that our conference did not break any paradigm of AI, did not find any groundbreaking solutions to AI problems and even did not unveil the mystery of consciousness.

Unlike most conferences, this one was not delimited and unified by a specific topic that attracts a particular coherent group of researchers. Instead of this, the conference participants were attracted rather by a shared personal attitude towards what they do—by their interest in what lies "beyond AI".

And so the conference hosted researchers with very different specialisations and scientific background. However, this kind of interdisciplinarity is always a risky enterprise: what if such an eclectic crowd turns into chaos where no one understands anyone else? But it did not happen this time. This time the participants—perhaps to their own surprise—realised that they shared a common language that allowed them to discuss, criticise, support, question or inspire their colleagues inhabiting often very distant areas of AI, ranging from AI engineering through philosophy to art.

This somehow reminds me of the legendary Macy Conferences held in New York between 1946–1953 where the giants of newborn cybernetics were gathering, everyone coming from so different fields only to find that they were all united by a common language—cyberspeak. Indeed, neither the scale nor "legendariness" on our side is what causes this elusive resemblance—rather it is this far-reaching mutual understanding which almost calls for a question: what common language lies beyond AI?

Pilsen, July 2012 Jan Romportl

Contents

Part I
Future of Artificial Intelligence

If a machine can create a better machine, will it do it? And if yes—and the intelligence explosion is born and path towards singularity open—what then? Should we be afraid of intelligence singularity, or should we look forward to it? Or simply do not care? Maybe we should rather be interested in what a single human-like mind is or should be capable of: analogy-making and aesthetic evaluation form a significant portion of our cognitive processes; wisdom is more than a formal knowledge system; a human-like mind is rational in a very specific way and at the same time it hosts a bottomless well of unconsciousness. All these issues should shape the future of artificial intelligence. And then we can ask: will a potentially emerging intelligence singularity be wise, with strong sense for analogies and aesthetics, humanly rational and still dwelling in realms of unconsciousness? If yes, then we should perhaps look forward to seeing its arrival. If not, then the word "intelligence" becomes hopelessly empty and we are left with the singularity as intelligent as a black hole, the Big Bang, or—to make it more fatal—even the Big Crunch.

Chapter 1
On the Way to Intelligence Singularity

Ivan M. Havel

Abstract. Since the fifties of the last century there have been debates about the so called "technological singularity", motivated by the predicted and later actual exponential growth of the speed and power of computers. Recently the interest of futurologists and philosophers shifts to the so called 'intelligence singularity' which some of them predict to happen soon after human intelligence is surpassed by artificial intelligence. This study critically analyzes certain assumptions behind the concept of intelligence singularity, in particular the idea of explosive growth of intelligence of machines with the ability of designing machines more intelligent than themselves.

1.1 What Is Singularity

It inseparably belongs to human nature that we are interested in our own personal future. This interest is necessarily interwoven with interest in our neighborhood—nearer, farther, up to the interest in world's future in general. But each of us perceives it differently. The interest usually dims with distance, not only in space but in time: tomorrow is of greater importance to me than the next month, year, decade—I was just about to write "and so on", but it would skip some difficultly expressible divide between what can be expected during our life and the life of our close ones on one hand, and what can happen in the far future on the other. It concerns the destiny of humankind, nature, planet, if not the entire universe. Personal interests diminish replaced by the vague care of the whole and this care eventually turns into curiosity.

I don't want to end at general banalities. I prefer to concentrate on one specific futurological topic—the less banal one, the more questionable. My concern in what follows is so-called *singularity*—let me explain what's going on.

Ivan M. Havel
Center for Theoretical Study, Charles University and the Academy of Sciences of the Czech Republic, Jilská 1, 110 00 Praha 1, Czech Republic
e-mail: havel@cts.cuni.cz

J. Kelemen et al. (Eds.): Beyond Artificial Intelligence, TIEI 4, pp. 3–26.
DOI: 10.1007/978-3-642-34422-0_1 © Springer-Verlag Berlin Heidelberg 2013

The term “singularity” is commonly used in the exact sciences (mathematics, physics, astronomy, cosmology, etc.) in various more or less precise meanings; in mathematics generally as a designation for the point at which a function acquires infinite or undefined value or where some equation has no solution. In astronomy it may be black holes, in cosmology the Big Bang or the Big Crunch. Meeting with the singularity usually requires a leap into another discourse, changing the existing way of thinking, arriving at something entirely new. As Zdeněk Neubauer writes: “The origin as a proper origin is a ‘singularity’—an occurrence or condition of such a nature (or structure) which cannot be described in the same conceptual framework that is applicable for its immediate surroundings.” [9]

Futurology, which thus indicates approaching a hypothetical limit of some accelerating process in the human world and in historical time, has borrowed the concept of singularity from the exact sciences. Futurological reflections (in the broad sense) allow the free play of imagination, which is a nice feature—because, besides other things, they mingle with science-fiction genre in an interesting way. My favorite example is the bizarre idea of an American physicist and essayist Alan Lightman, who in his charming book *Einstein’s Dreams* [8] imagined a hypothetical world in which everybody knows the day and minute of the end of time. One day before the end people indulge in unbridled laughter. Then, in the very last minute everyone falls silent; people hold hands and form a giant circle. As if they leaped off a peak and the bottom of the great deep hurtle nearer and nearer—they all share the same fate.

A certain specific type of singularity, sometimes called *the technological singularity*, became to be discussed within the context of a rapidly accelerating development of computing and information technology. The main inspiration is often assigned to Moore’s Law, according to which every two years the density of transistors in integrated circuits doubles.[1] Similar estimates exist for various other technological growths, such as the increasing number of elementary operations per unit time (measured in billions per second nowadays!) or drastic reduction of energy requirements. Here we simply note the one thing in common: the value of some quantity within a fixed time interval always *multiplies* (by a constant greater than 1), leading to its *exponential growth* over time. After all we are living witnesses of such growth—doesn’t it seem to you that the time lapse between new surprises on the IT market is being more and more reduced? Who would guess a few years ago, how today’s smart phones and tablets, web search engines and positioning systems would look like?

In such considerations some people fall to futuristic euphoria, others, on the contrary, frighten us by various catastrophic scenarios; everyone is on edge. Futurology as a science should be particularly interested in whether some technological singularity will actually occur, when it would happen, how it would treat us and how we would treat it. However, concrete ideas about the nature and variants of technological singularities are either very vague or too focused on one or another professional aspect. There is however a common denominator: today’s technological development is not only *fast*, it is also *accelerated* and, even more, it is *accelerating itself.*

[1] See, e.g. `http://www.intel.com/technology/mooreslaw/`. Moore’s estimate (from 1965) was later generalized in a different way and updated according to the actual development.

And where a self-acceleration occurs, there is an acceleration of acceleration, an acceleration of acceleration of acceleration... until it gets out of joints and ends in a giant explosion called a singularity.

In a similar sense the well-known mathematician Stanislaw Ulam might have understood the term "singularity" already in 1958, when he reported on his conversation with John von Neumann about "... the ever accelerating progress of technology and changes in the mode of human life, which gives the appearance of approaching some essential singularity in the history of the race beyond which human affairs, as we know them, could not continue. " [12]

Technological development undoubtedly plays an important role, since it provides incredible technical achievements from extreme miniaturization through a huge memory capacity and computational speed up to the use of new physical and biologically motivated principles. Not speaking of information networks like the Internet, the Web, and other global or perhaps even satellite systems. But aren't they all just aids to something even more significant? What about talking about *intelligence* in a deeper sense? Surely, from our normal, *human* intelligence we cannot expect any precipitous development. But there is already a maturing *artificial intelligence*. Just this intelligence, with its hidden and perhaps unsuspected possibility of radical, up to avalanche-like increase, caused the recent reflections of a particular type of singularity. We will call it the *intelligence singularity* or shortly *Singularity* (with capital S).

1.2 Artificial Intelligence Ready to Start

Firstly, a few words about *artificial intelligence* itself. Today, this set phrase refers rather to one of the academic, research, and to some extent programming disciplines within informatics or computer science, but it seems that the time has come to begin to understand both the words again in their literal meaning,[2] so that they relate to something that is for one thing *artificial* and for another *intelligent* and that may *soon* really *occur* among us (in variably vague meanings of all four words in italics).[3] Here, speculative thinkers may already pick up the baton from cyberneticists and computer scientists with full engagement of free fantasy, even vivid one—many already do, among them technically and application oriented researchers with many years of professional experience.[4]

Let's try to run a bit with them. It means: (a) to imagine that artificial intelligence (in the literal meaning) has already been implemented (or at least appears to be

[2] Once the phrase "artificial intelligence" was indeed taken literally, leading to multiple misunderstandings of a philosophical nature. However, its use for a scientific discipline later prevailed.

[3] Without obligation we can imagine such an implementation in a computer, in an algorithmic system, an information network, or with the help of a mechanical copy of the human brain, neuron by neuron, synapsis by synapsis (called brain emulation). The meaning of the word "intelligent" will be discussed further on.

[4] There are also institutions for that, including The Singularity Institute for Artificial Intelligence (`singinst.org`) or Future of Humanity Institute (`www.fhi.ox.ac.uk`).

feasible) at the level of human intelligence, (b) to think about the possibility (or necessity) of intelligence explosion (i.e. extremely rapid growth) and (c) to ask whether this explosion can (or must) have the character of the Singularity in finite time. (I am phrasing it carefully to avoid the impression that all I write I take for real. My point is rather to explore the possible arguments for and against the idea of Singularity—whether as a concept, as a phenomenon, or as an event in the near future).

The concept of the intelligence explosion is often presented in already a classical text by I. J. Good from 1965 (the very term Singularity, in our sense, has appeared only later—see for example [13],[15],[7]):

> Let an ultraintelligent machine be defined as a machine that can far surpass all the intellectual activities of any man however clever. Since the design of machines is one of these intellectual activities, an ultra-intelligent machine could design even better machines; there would then unquestionably be an "intelligence explosion," and the intelligence of man would be left far behind. Thus the first ultraintelligent machine is the last invention that man need ever make. [4]

The formulation is seemingly quite understandable and not lacking a logic, after all it has remained stable only in small variations for the last 50 years. Let me quote its latest version from a recent study by well-known philosopher of mind David Chalmers:

> The key idea is that a machine that is more intelligent than humans will be better than humans at designing machines. So it will be capable of designing a machine more intelligent than the most intelligent machine that humans can design. So if it is itself designed by humans, it will be capable of designing a machine more intelligent than itself. By similar reasoning, this next machine will also be capable of designing a machine more intelligent than itself. If every machine in turn does what it is capable of, we should expect a sequence of ever more intelligent machines.[1, pp. 7–8]

Chalmers' formulation is (at least it seems to me so) somewhat clearer and logically more accurate than Good's, but they both express the same basic idea. So I'll just continue to advert to these two—almost canonical—presentations of nowadays widely accepted hypothesis about the genesis of Singularity as to (two) *"representative theses"*. They will save us from dealing with dozens praiseworthy works of other authors on a similar theme; as it will turn out, already these two theses (even one of them would be enough) provide more than enough ideas to ponder.

Indeed: already at the first reading we will notice that both representative theses are based on certain important tacit assumptions. And just these tacit assumptions evoke deeper (or wider) issues, which I want to deal with in the following sections. Let us list some of the issues already here:

- Are we refering to our real future, or to a fictitious possible world?
- The word "intelligence"—what does it actually mean? Who (or what) is its bearer or performer?
- Is it possible to compare various qualities of intelligence, improve them, and perhaps even measure them? In other words, how to understand the terms "surpass", "better", "more intelligent", etc. in our representative theses?

- Is also the capability of "designing an intelligent machine" a part of the intelligence of a machine? Even the capability of "designing a machine more intelligent than itself"?
- How does the intelligence explosion proceed and what is the nature of explosiveness of such an explosion?
- If a machine were able to create a better machine, would it really do it?

Neither of our two representative theses directly mentions a (general) singularity, nor even the (intelligence) Singularity. We could therefore become interested in whether and how the concept of singularity could relate to the concept of explosion. These two concepts are not clearly distinguished in Chalmers' paper; so I make the distinction myself: (in general) I will view explosion as a *temporal process* of certain kind, and singularity as an *event* that might happen sometimes in future (typically due to a rapid explosion). Therefore we may ask further questions:

- What type of explosion ends up in a singularity? Will an intelligence explosion result in the Singularity?
- If it does, what will happen then?

1.3 The Dual Nature of the Future Tense

Note that in our two representative theses it is not clear to what extent the authors play a free game in a hypothetically conceivable "possible world", and to what extent they have on mind something that could "really" happen in our present actual world. There is a simple origin of difficulty of such distinction: in *both* cases we deal with mental constructs. The only difference is that in the first case our considerations are (perhaps implicitly) related to the abstract, imaginary time, while in the second case they are intended as related to the anticipated earthly time (that once will turn into "historical time"). The grammatical future tense of the verbs alone cannot distinguish it.

I'm mentioning it just because our topic is, by its very nature, ultimately futurological: will there be a Singularity, or not? There is always the possibility that it may occur in "our" actual time. The context, in which one or another statement appears, or the knowledge of the author's interests would naturally help to distinguish the real earthly time from the imaginary one; in this paper, however, we will avoid this problem quite easily: as far as a conceptual analysis is concerned, the reality of time is irrelevant.

In fact, debates about various estimations of *when* a Singularity really comes prevail. (For example, one of close estimates is around 2035—so that younger readers will be still around![5]) For some reason people like to read just about *our* near future, it is an attractive topic, indeed.

Some people, however, prefer to play with conceivability in principle (let's remember the Lightman's fiction mentioned above), others declare themselves as

[5] I recommend Google: Singularity 2035.

futuristic visionaries and foretell either brilliant or disastrous futures. Some others don't make such a distinction.

1.4 Intelligence—What Does It Actually Mean?

In our representative theses the occurrences of the word "intelligence" are mostly wreathed in comparative or even superlative modifiers ("more intelligent", "the most intelligent", "ultra-intelligent", "intelligence of man")—and so what shall we think of *intelligence as such*? Can it be viewed as a capability, or potentiality, or skill, or a gift? Alternatively, as an intricately interconnected complex of many different *component skills*, abilities and potentialities? Or does it make sense to talk about intelligence only after we have a clear idea of *who* (or *what*) is its bearer, owner or user, at least in terms of his (her, its) genus (human, dolphin, mouse, computer, cellular phone) and in that case also roughly *when* we encounter it (in the Holocene, today, soon, in the next century)?

Our representative theses presume, in fact, just four general characteristics of intelligence, namely

1. That it can be *improved*,
2. That it can be attributed to *both* humans and machines,
3. That machine intelligence can even *surpass* human intelligence, and finally,
4. That it somehow includes the *capability of designing* something (namely other intelligent machines).

These characteristics *per se* would not help us to define the concept of intelligence (the first three of them could apply to many other attributes, such as hardness, noisiness, vibration, etc.). We can only rely on a natural, intuitive view. When I think that someone is smarter than I, I usually do not investigate how I mean it. I just somehow see it. Someone may naturally disagree with me: I, for instance, may appreciate speculative thinking, while the other appraises solid evidence—two attitudes that are almost in the opposite; at least for a single individual.

It is therefore useful to introduce a certain conceptual differentiation. In our representative theses intelligence is not understood in the way psychologists see it, i.e. as a faculty attributed to a concrete individual (who is typically either a person or even a machine or other entity, depending on the case). It is rather an imaginary aggregate of "all the best" aptitudes within a given group; to distinguish it from individual intelligence I will call it the *generic intelligence*. Thus we say "human intelligence" or equivalently "the intelligence of man" and we understand it well, albeit with the amount of uncertainty.[6] Correspondingly, I will introduce the concept of a *generic bearer*—an imaginary representative of the given group—whose individual intelligence equals generic intelligence of the group (it does not have to be the individual intelligence of any member of the group). For the generic bearer we can thus allow for coexistence of even those properties that would be mutually incompatible for an individual bearer. In this way it is possible to understand the

[6] For example the claim that a man is able to prove Fermat's Last Theorem is correct.

reference to "however clever" man (Good) and formulations about humans in plural, or about a machine in singular (Chalmers).[7]

Unless the context indicates otherwise, in the following I will use the term intelligence for generic intelligence.

Let us now return to the four above-mentioned general characteristics of intelligence.

Ad 1. The first characteristic assumes that intelligence *can be improved.* It is worth noting that the very term "improve" reveals a certain positive attitude (perhaps typical for experts), but let us leave the question of valuation aside.

Initially we do not have to provide the term "improvement" with any precise meaning. Rather we may refer to the fact that we already have an elementary idea—whether intuitive (for a human) or technically well-founded (for a machine)—that some components of intelligence can, at least minimally, be improved (e.g. the extension and reliability of memory, speed of decision making, the range of the hierarchy of logical levels, the ability to learn, discriminability of details, etc.). Well, if *minimally*, why not a little *more*? When *little* more, why not *much* more? I remind that it is only a preliminary intuition, in the next step we would have to be engaged in more detail in quantification or direct measurement of intelligence. Let's postpone it to the next section.

Ad 2. The second characteristic opens a serious issue: whether it is possible to talk about both human and machine intelligence in one breath, as it were, and moreover so that they can be measured on a common scale—or at least compared as for which of them is "better" or "superior." This is suggested by Good's phrases "a machine that can far surpass all the intellectual activities of any man however clever" and "the intelligence of man would be left far behind" or Chalmers' phrase "a machine that is more intelligent than humans". If we are not content with mere comparisons of performance—indeed, still popular (from the Turing's test to chess tournaments)—we quickly encounter the problem of different specific dimensions of intelligence for humans and machines.[8] What is even more important is that

[7] The concept of generic intelligence has to be distinguished from that of *collective intelligence.* The latter may be viewed not as a result of mere addition or aggregation of individual intelligences, but rather as a higher-order emergent phenomenon based on a complex network of interactions between individuals in a group. Then we can view the whole group together with the network of interactions as a higher-order individual, which is a single bearer of *its own* individual intelligence. Thus we do not need to consider collective intelligence as a specific new category of intelligence. As a matter of fact, we can understand the collective intelligence of mankind as a case of individual intelligence, provided we view mankind as a higher-order individual. (Perhaps, one try to extend the idea downwards and regard the intelligence of any individual human as a collective intelligence of neurons or neuronal clusters in the brain).

[8] Occasional coincidence of names for certain dimensions reveals little, since in the case of machines such names are often metaphorical allusions to the human qualities. Differences between humans and machines is a permanently debated issue (let me remind just the distinction between "brute force" and "heuristic procedures" in solving combinatorial tasks).

intelligence may be associated with other qualities that can be ascribed only to humans, or only to machines. For humans it may be for instance intuition, ingenuity, inventiveness, imagination, empathy, thinking in images, metaphoric and analogical reasoning, vague thinking and many other faculties, not mentioning subjective feelings and experiences. For machines it may be, say, a tangled hierarchy of algorithmic and logical levels, each involving a strong parallelism, and ultimately the celebrated unhuman repetitive patience and tirelessness.

Ad 3. The third characteristic then does not actually bring anything surprising. It is quite obvious that even if we anticipated some significant improvements to human (generic) intelligence, for example considering various brain stimulations, meditative practices and extraordinary savant skills, we will soon hit against various biological, physical, and other limitations. Of course, machines have their limitations too, mostly physical; these are still relatively beyond the reach of current practice and one can always expect some unpredictable technological and design innovations. It is therefore not surprising that speculations about an intelligence explosion are focusing mainly on *artificial* intelligence, even in the broadest and quite flexible sense of the word.

Ad 4. Finally, the fourth characteristic is slightly more concrete, rendering intelligence in connection with the ability of designing machines, even intelligent machines. This connection can be understood in two ways: we can regard this ability either as one of the genuine dimensions of intelligence, or just as a side-effect that intelligence only supports. I will restrict myself to the first case. Provided intelligence is understood in the generic sense, it may include, by definition, capabilities that are not shared by *all* members of a given group. Hence we have no problem with the fact that not every human is able to design machines (not talking of intelligent machines).

We arrive at the notion of intelligence as a large aggregate of various skills, capabilities and talents. A lot has been written[9] about them—at least in the case of *human* intelligence, I do not intend to recapitulate it all here since for our theme it is not important. Being aware of the difficulties with the comparability (see above), I will also skip the question whether machines (especially computers), being human artifacts, can in one or another dimension of intelligence in principle equal humans, or perhaps even surpass them, at least a little (the discussion about this issue has already lasted for more than half a century and I do not know how I could contribute to it at the moment).

However, to proceed, we have to agree at least on something: we will take up a *positive answer* to the just mentioned question (whether machines can equal humans), and even its more ambitious version (whether machines can surpass humans, at least a little), as a purely *hypothetical assumption* and not as an empirical or

[9] See for example Gardner's classification of nine dimensions of human intelligence [2].

logical fact.[10] This assumption will simplify our further considerations at least to the extent that we may forget about humans and focus directly on (generic) *artificial* intelligence. I will dare to call the generic bearer of this intelligence *the* machine (as the imaginary ideal representative of all realizations of artificial intelligence at a certain stage of development).

The classical AI research was usually focused on selected particular faculties, which (in humans) are counted as intelligent, and tried these faculties—to a certain extent successfully—implement (on a computer). We know them well: chess playing, pattern recognition, problem solving, theorem proving, speech processing, etc. The very name of the discipline implies that all these particular faculties could be regarded as specific manifestations of a single potentiality, quality or power. Some call it *general intelligence*.[11] Perhaps it would be difficult to somehow define general intelligence; we know only those specific manifestations and even those cannot be summarized into a comprehensive list. It is rather a conceptual construct without direct ontological support.[12]

The idea of general intelligence can be associated with a certain orientation of the AI research—luckily we can talk of orientation even under a relatively vague concept of a goal. Marvin Minsky, one of the founders and leading representatives of the discipline, advocated such an orientation already a decade ago:

> Only a small community has concentrated on general intelligence. No one has tried to make a thinking machine and then teach it chess—or the very sophisticated oriental board game Go [...] The bottom line is that we really haven't progressed too far toward a truly intelligent machine. We have collections of dumb specialists in small domains; the true majesty of general intelligence still awaits our attack. [...] We have got to get back to the deepest questions of AI and general intelligence and quit wasting time on little projects that don't contribute to the main goal. [11]

Today, the artificial general intelligence (as a research project) is being pursued,[13] but it still waits for clarification whether and how general intelligence of machines relates to that of humans. This is another reason for caution when talking in one

---

[10] I emphasize that on my part it is really just a working hypothesis, not faith, such as in V. Vinge who almost 20 years ago wrote: "... I believe that the creation of greater than human intelligence will occur during the next thirty years." [13] I will also not comment Chalmers' arguments in favour of this hypothesis (they happened not to convince me). The mentioned hypothesis actually claims the possibility of so-called *strong artificial intelligence* (strong AI, the term of John Searle).

[11] It is good to note again a conceptual distinction, this time between general intelligence and generic intelligence: in the former case we talk about unification of a set of subcomponents of "one" intelligence, in the latter case about the unification (or aggregation) of a set of individual intelligences. These two types of unification are complementary.

[12] There is a certain analogy with the concept of energy in physics (intuitively understood). We have only indirect experience with energy through its various manifestations; in contrast to the intelligence we are able to formalize energy by means of mathematical tools.

[13] The project already exists and has its abbreviation AGI (artificial general intelligence). See for example [3].

breath of intelligence of machines and humans. This is linked to the theme of the next section.

1.5 Can Intelligence Be Measured?

Variable quantities are significant part of the description of any explosion. How else could we talk, e.g., about rising pressure (in detonations), energy release (in nuclear chain reactions), energy density (in gravitational collapses), anxiety (in panic attacks), population growth, etc.?

The same holds for the presumed intelligence explosion, albeit intelligence is much more ambiguous, vague, and context-sensitive phenomenon. If intelligence is to be a driving force of an explosion, we cannot avoid its, at least partial, quantification. Recall terms used by Good or Chalmers: "surpass", "however clever", "better", "more intelligent", "ever more intelligent", and "most intelligent". Moreover, these terms are used once in the context of machines, once of humans, usually of both. How to understand them?

In an attempt to somehow evaluate the degree of intelligence in terms of intensity, applicability, importance etc. (not to mention its dependence on other properties) we will at first probably try to use some scalar quantity, preferably such that its values are integers or real numbers. As known, attempts to measure individual human intelligence are known for hundred years,[14] but they are restricted to the outside, easily testable *manifestations* of certain selected abilities, and indirectly only to the *dispositions* to them, while no regard is taken to the interconnection of such manifestations, the less to possible negative correlations among them. The obstacle is the essential impossibility to measure qualities like imagination, creativity, inventiveness, intuition, mental flexibility, etc. One can only guess a countless number of other abilities (if we can even talk about a number), that not only elude scientific investigation but may not be yet discovered. And even if we knew about them and could endow them with some sort of a "measure", there is no hope to identify corresponding abilities of machines. In fact, such incomprehensible qualities may be crucial for a design of intelligent machines. Not counting the understanding of the phenomenon of intelligence as such!

Fortunately we are not in a situation that would require an exact approach. In cases of very general, theoretical and mostly speculative considerations (others are not at issue here), intuitive assumptions that this or that is simply *conceivable* is perfectly sufficient. We do not need to transform everything into numbers, because we do not want to create exact mathematical models, prognoses and statistical estimates. (An explosive does not need to compute or measure something about itself in order to know how to properly explode.) Nothing prevents us from grounding our theoretical considerations about intelligence explosion on freely chosen analogies and "soft" metaphors.

[14] Intelligence quotient (IQ) was defined in 1912 by German psychologist W. Stern; later many variants of intelligence measures were proposed.

Somewhat paradoxically we may credulously rely even on analogies and metaphors borrowed from mathematics. Here I will tentatively rely on the imaginary[15] idea that all conceivable cases of intelligence (of people, machines, whatever) are represented by points in a certain abstract multi-dimensional "super space" that I will call the *intelligence space* (shortly IS).[16]

Imagine that a specific coordinate axis in IS is assigned to any *conceivable* particular ability, whether human, machine, shared, or unknown (all axes having one common origin).[17] If the ability is measurable the assigned axis is endowed with a corresponding scale. Hypothetically, we can also assign scalar axes to abilities, for which only relations like "weaker-stronger", "better-worse", "less-more" etc. are meaningful; finally, abilities that may be only present or absent may be assigned with "axes" of two (logical) values (*yes-no*). Let us assume that all coordinate axes are oriented in such a way that greater distance from the common origin always corresponds to larger extent, higher grade, or at least to the presence of the corresponding ability.

The idea is that for each individual intelligence (i.e. the intelligence of a particular person, machine, network, etc.), as well as for each generic intelligence (of some group) there exists just one *representing point* in IS, whose coordinates determine the extent of involvement of particular abilities.[18] As we shall see, the concept of IS will ease the "visualization" of various degrees of intelligence, differences between different intelligences, and in particular, the development (e.g. the growth) of intelligence over time (which is the topic of the next section). In essence, the aim is to help natural intuition with simple geometric metaphor. (Note that it is just a metaphorical visualization, by no means a mathematical model!)

It is important to realize that due to qualitative differences and thus mutual incomparability of different concrete abilities, the exact distance of the representing point from the origin in IS does not comply with anything real, and so it would not help in establishing any *absolute intelligence measure* on which the comparative relations such as "more intelligent", "better", etc. would be based.[19] There is fortunately another, albeit partial solution: first, to distinguish different *types* of intelligence, characterized by relative participation (i.e., relative "weight") of specific abilities. I will call each such structure of relative participations a *profile* of intelligence. Without going into details we can imagine that the profile of intelligence

---

[15] Further on I will often skip the attribute "imaginary"—*everything* will be imaginary in some sense!

[16] If we take an occurrence of intelligence as a state, IS can be regarded as a kind of a state space.

[17] Whoever wants to imagine IS visually is advised to choose the usual Cartesian space provided he doesn't feel too restricted by its three dimensions.

[18] The location of this point (or its coordinates) in the case of generic intelligence of some group depends not only on the individual intelligences of its members, but also on the way how generic intelligence derives from them. (Example: each coordinate is maximized over all members of the group at representing point for the generic intelligence of a group in which there is an effective "division of labour".)

[19] In this respect, the geometric intuition of Cartesian coordinates with the usual metric fails.

characterizes certain rough "direction" in IS. This may be a line passing through the origin, or a neighborhood of such a line, or even a whole subspace of IS (delimited only by the abilities relevant to a certain type of intelligence).[20] Only when two points lie roughly in the same direction, their distance can serve as a *measure of the difference* between relevant intelligences. On the other hand, when individual intelligences have different profiles (different directions in IS), we cannot compare them. Profiles (directions) that could be called *orthogonal* are the extreme case of the difference. I have already indicated an example: someone excels in logically precise judgments but is poor in imagination and speculative thinking; somebody else is exactly the opposite. These two types of intelligence may even suppress each other (they co-exist only in sufficiently universal generic intelligence).

Though our IS appears to be somewhat vague and abstract it offers a rather universal common framework for partial intelligence measures, and even for diversification of types (profiles) of intelligence at high level of generality (human, machine, animal, hypothetical) and their various typological refinements.[21] For example, the general type (or profile, direction in IS) called "human being" may include subtypes "educated European", "chess master", "numerical savant", "futurologist" etc. (here each term in quotation marks points to the generic representative of the corresponding group.) It is obvious that the more specific a profile, the easier it is to compare various levels of intelligence within the profile. Remember, however, that still we are playing with imaginary concepts and relations, while the real, empirical concretization of similar typologies would generally fail for various reasons, including ethical ones.[22]

1.6 Intelligence in Motion

So far we have dealt with the intelligence space only in static terms. But once we have some idea, albeit vague, about the measure for intelligence of a given type (profile), we can be concerned with the temporal process that leads to the improvement of the intelligence in question, in other words to gradual *increase* of its intelligence measure. Otherwise it would be impossible to talk about intelligence explosion or about the Singularity. For this purpose it is necessary to enrich the concept of IS by considering movement in it.

[20] This general formulation avoids considering real types of intelligence (for example, those dealt with by cognitive psychology). Only very few "directions" in IS, or intelligence profiles, could correspond to something real, but nobody knows what to expect from the future development of artificial intelligence.

[21] There is still another aspect which I do not take into consideration here. In IS we can represent a given type of intelligence not only by a certain direction but also by a *horizon*. The horizon (of a type of intelligence) can be thought of as a limit of distance from the origin that for some or another reason cannot be surpassed (this applies typically to the physical bearers of intelligence). In fact, the hypothesis of intelligence explosion presumes the possibility of "pushing" the horizon further away in a specific direction.

[22] Let us recall the fuss about James Watson's statement about "otherness" of the intelligence of African black people.

The basic intuition is simple. In our approach, an individual intelligence (its profile and measure) projects onto a hypothetical representing point in IS, so whenever there's a change of intelligence, this change is reflected in the movement of the representing point. It suffices to focus only on the movement along a certain trajectory in IS.[23] Thanks to it we do not have to deal with a variety of technical aspects—for instance how the underlying technology and functional principles are changing or whether entirely new machine realizations emerge—even though we should not forget that just these aspects influence the development of intelligence.

Let us first notice several fundamental types of movement of a representing point in IS:

1. Receding (in a given direction) $\rightarrow$ increase of intelligence without change of its profile;
2. Turning (to another direction) $\rightarrow$ change of the intelligence profile, e.g. "retraining";
3. Occupying additional dimension of IS $\rightarrow$ discovery of a new relevant skill, hence extension of the profile;
4. Local "wavering" $\rightarrow$ minor fluctuations in level and profile of intelligence (depending on its actual bearers).

In our context, the key role is played by receding, so I will limit myself to 1 (generally we can expect a combination of 1 and 2, whereas 3 can be considered as a special case of 2; and we can happily ignore local wavering 4).

But what cannot be ignored is *time* (whether real or abstract, which is yet to be distinguished). If the distance of the representing point (from the origin of IS) represents the intelligence measure[24], we can imagine increase of intelligence, i.e. a growth of its measure, as a gradual receding of this point from the origin. This movement may accelerate,[25] and when even the acceleration itself also accelerates, there is a precipitous "escape" of the representing point towards infinity. It is the explosion. If the escape is "completed" already in a finite time, we have the Singularity.

However, time is not directly displayed in IS, at least if trajectories are not tagged with temporal marks. For various reasons it is preferable to illustrate temporal dependencies graphically: the horizontal axis for time, the vertical axis (in our case) for the intelligence measure. Then, for example, a statement that "an intelligence

[23] It would be nice if we could examine (and thus predict) the developmental changes in the typological profile of intelligence, various new trends, and newly discovered dimensions of intelligence, and how they can (or will) be reflected in the shape of the trajectory. Vagueness of our approach does not allow it, except for aforementioned distinction of several types of growth of intelligence measure, represented by the distance of the representing point from the origin.

[24] Let us assume that everything needed for the existence of this measure is fulfilled (it might not be!).

[25] It may, however, slow down and eventually converge to some finite limit (a horizon) or simply stop. Here, I do not consider such cases—however they are common in the real world, they are not directly related to our topic.

grows” obtains a convenient graphic form that shows even *how* it grows (for simplicity I do not take into account here possible changes of intelligence profile[26]).

Generally speaking, we can talk about an explosion whenever some quantity is not only growing, but the speed of the growth is growing as well. A typical example is an exponential function of time; mathematically it is defined for all values of its argument (i.e. time), so that no singularity may occur *in finite* time. However, this may happen (for example) in the case of a hyperbolic function of time. Only then the term “singularity” is formally appropriate. Therefore I distinguish the (more general) concept of an explosion from the (more radical) concept of singularity.[27]

As mentioned, in our case the (increasing) function of time is the chosen intelligence measure. I suggested how we could imagine such a measure, but only for particular type (profile) of intelligence and ignoring the real natural or physical causes of its progression. At our level of vagueness, however, it is not appropriate to categorize different types of growth with the help of exact mathematical models.[28] Hence what is only left for theoretical treatment of Singularity is its *conceivability*.

1.7 How the Explosion Is Born

With the above conclusion I do not want to digress from the subject. The fundamental question, what *in principle* can start and nourish a hypothetical explosion,[29] is still intact. We have no direct experience with intelligence explosion (much less with the Singularity indeed), nor we can offer specific arguments in favor of the hypothesis that artificial intelligence can overcome human intelligence. The only observable phenomenon today is the accelerating development of technology—but it only demonstrates a technological explosion rather than a proper intelligence explosion. What remains is to disregard the current state of technology and rather to

[26] This restriction is not trivial—we cannot represent real intelligence by a one-dimensional measure that could be simply plotted above a linear axis. This would correspond to a fixed intelligence profile (fixed direction in IS) and therefore only *one-directional* development of intelligence and *one-directional* explosion. From the futurological viewpoint this would be rather serious limitation (we wouldn’t get more than an analogy to Moore’s Law). In general, it is necessary also to take into account changes of the intelligence *profile*, which may depend, for example, on a decline of one ability compensated by appearence of a different ability. Probably it would not cause a serious problem for the graphical representation of the temporal plot; however, it would be easier to limit ourselves to the one-directional case.

[27] I actually do so to please the futurologists who prefer discussing events that will not miss us.

[28] A. Sandberg [10] compares different mathematical models of accelerated growth in different areas (technological, scientific, economic, biological, population development etc.). For our study, however, Sandberg’s sorting does not have greater than heuristic importance.

[29] Or what can slow it down or stop it—but this is not the point here.

concentrate—in view of the phrase "in principle"—on semantic, logical, and (as it will turn out) cybernetical aspects of the problem.[30]

Let us recall one part of Chalmers's thesis, where he claims that "a machine that is more intelligent than humans will be better than humans at designing machines". And he continues, "So it will be capable of designing a machine more intelligent than the most intelligent machine that humans can design." It seems that it is quite sensible deduction, but let us note that it itself presupposes something not quite obvious, namely (a) that a more intelligent machine is (also) better at designing machines, and at the same time (b) if a machine is better at designing machines, then it is capable to design better (more intelligent) machines. I will return to the point (a) at the end of the section; first I'd like to hint at a catch in statement (b).

Is it really true, generally speaking, that a better producer produces better products? After all, a better clockmaker may not make better clocks, and a better fisherman may not fish better fish (rather mischievous counterexample). In common speech we mostly ignore this delicate distinction (indeed, a better cook does cook better meals and a better painter does paint better paintings), but in our case we should be more cautious and preferably postulate the statement of (b) explicitly (i.e. to include it already in the definition of the relation "to be better at designing").

Several other concepts would deserve a similar caution, but I'd rather get to the conditions of intelligence explosion as such.

The main argument for its outset and progress significantly reminds the general principle of *positive feedback* in cybernetics.[31] We can demonstrate it—in a simplified way and in the form suitable for us—by an example of the feedback amplifier of some physical quantity. The behavior of the amplifier can be described in three phases: (1) an initial value of the quantity is brought to the input of the amplifier, (2) the amplifier increases the value (in a way that can be expressed by an appropriate transfer function)[32] and (3) the so increased value is brought *back* from the output to the input of the same amplifier (hence the term "feedback loop"). This cycle is repeated unlimited number of times, and if nothing stopped it, the value of the quantity would theoretically grow beyond any limit. In real systems too rapid growth usually leads to an overload, oscillation, overheating, satiation, overgrowth, explosion, or alternatively to a collapse. Such events usually preclude further growth.

In the intelligence explosion, however, we have nothing to do with the usual physical quantity, but with the imaginary and vaguely defined measure of intelligence. Is it even then possible to base the hypothesis of explosive growth on the analogy with the principle of feedback? It is just enough to read the theses of Good and Chalmers to notice that they both are based on a certain general principle, let us call it

[30] Let the reader be prepared for somewhat technical nature of some of the arguments in this chapter.

[31] The very term "feedback" is older; cyberneticists of the middle of the last century has elevated it to a general principle which is applied in many diverse fields, most often in the form of *negative* (stabilizing) feedback.

[32] It can also reduce or change it otherwise, but it does not need to concern us.

self-relatedness.[33] Since feedback is based on the same principle, mutual comparison may help to understand the nature of intelligence explosion.

The representative theses of Good and Chalmers can be rephrased in the form of the following two claims:

1. *The ability to improve something is used to improve the very same ability.*
2. *The machine has the ability to design a better machine than itself, including the fact that the new machine has the same ability again, i.e. the ability to design a better machine than itself.*

The first statement is somewhat more abstract; the second seems to be its special case. However, in both claims there is something implicit—and just this may be crucial for the emergence of intelligence explosion or Singularity.

Let us consider claim 1. Do we understand it correctly? The first problem is already the meaning of the indefinite pronoun "something"—does it mean "anything"? That would be overly useful ability, indeed, though it is hard to imagine how it could be realized—already because the meaning of the word "improve" crucially depends on the nature of *whatever* is being improved. What about to consider the opposite extreme, namely, to maximally reduce the list of candidates for improvement. In fact, claim 1 assumes merely one candidate, see the ability to improve. But again: what to improve? Well, itself. So we are caught in a strange loop.

Maybe there's a compromise solution: to associate with every "ability to improve something" a list of precisely those things that can be improved thanks to the said ability. A question arises, of course, whether a change of this ability, like its improving, would not affect the list. After all, isn't an *enrichment* of such a list also an *improvement* of the improving ability? If the explosion is applied just in this respect, it would lead us again towards that overly useful ability to improve anything!

Fortunately, we have the analogy with amplifiers (amplifiers exist in real world!). For an amplifier, the analogy of claim 1 would have more specific form:

3. *The ability of the amplifier to increase the value of a quantity is used to increase the value acquired by the (previous) use of the same ability.*.[34]

There is no problem with this. The formal difference between "improving ability" and "improved quantity" and explicit reference to time (note the word "previous") obviously helped. I will return to the issue of time in the next section.

We still have claim 2, which involves another distinction: it talks partly about the *ability* to improve something and partly about the *machine* which is the bearer as well as the object of that ability. There is, however, a new ambiguity in the expression "including the fact": the word "including" either refers to "design" or even to

[33] As self-related we may regard various phenomena and processes that affect themselves (e.g. their own behavior) in a certain way, the feedback loop being a special case. The concepts of self-relatedness and self-reference are often conflated, even if the latter is a special case of the former.

[34] Some may prefer a symbolic version: The ability A of an amplifier to increase, for some quantity Y, its value $Y(t)$ to $Y(t') > Y(t)$ is used to increase the value of $Y(t')$ (acquired by previous use of the ability A) to $Y(t'') > Y(t')$.

"better machine". In the first case we would not expect more than a simple transfer of the said ability to the new machine, in the latter case, moreover, we would presume an improvement of the transferred ability as such. Although it is a similar problem as in claim 1, it may be useful to phrase it differently.

Above-mentioned ambiguity of the word "including" cannot be removed without making clear what is generally meant by the expression "better machine"—that either (a) the machine is better *in everything* (and thus also in the ability in question), or (b) it is better *only in something* (perhaps without being worse in anything else). It seems that option (a) more likely leads to an explosion (but it makes harder to quantify it); while option (b) could lead to the explosion only if a particular ability, namely the ability to design a better machine, is among those under improvement. In fact, it could be the only one that is improved—but it is not guaranteed that its change would not preclude any further improvement.[35]

I do not think the reader should follow this logical analysis; it is enough to realize how the used concepts may be sensitive to various tacit assumptions, especially when a self-relatedness is in play.

In the case of claim 2, its version for amplifier may help again:

4. *In the amplifier a quantity can obtain a higher value, and—thanks to a positive feedback—in the same amplifier the quantity can obtain even higher value.*[36]

I do not want to play with words any more, I only want to make a comparison: on one hand we have somewhat opaque claims 1 and 2 referring to the phenomenon of self-relatedness for improving machines (hence intelligence), on the other hand, there are analogous and somewhat clearer claims 3 and 4 about amplifiers with positive feedback. Let us note the important difference: in the former case the improvement required creating always something new (a better ability, a better machine), and therefore it was necessary to guarantee an "inheritance" of certain qualities or abilities. In contrast, in the latter case we have always one and the same quantity that only changes its values.

From this comparison we can get an idea of some conceptual shift. I will indicate it in a simplified way for the case of artificial intelligence and its development.

The representative theses of Good and Chalmers, from which we started, induce the idea of a discrete series of machines (I remind the generic meaning of the word "machine"), in which every subsequent machine is more sophisticated (with increasing intelligence level) than the preceding machine. Now let us change a perspective and instead of such a series of machines let us think only of *one single machine* that gradually improves itself (either continuously or in steps).[37] Such a perspective is indeed completely natural for our concept of variable measure and profile of intelligence, as well as for the concept of generic bearer of intelligence with "his" unique trajectory in IS. Instead of improvement and perfection we can now

[35] Designing machines is a very complex process, so that its possible improvement may significantly change the intelligence profile.

[36] Again with symbols: In the amplifier a quantity Y can obtain value $Y(t') > Y(t)$, and—thanks to a positive feedback—in the same amplifier it can obtain the value $Y(t'') > Y(t')$.

[37] In other words, the identity of species transforms into the identity of the individual.

talk about *self-improvement* and *self-perfection*. The problem with inheritance disappears, similarly as it disappears in the case of feedback amplifier.

At the beginning of this chapter two postulates (a) and (b) were mentioned. Now, the first of them (namely that among various partial components of intelligence is the ability to design machines) transforms into a rather more natural assumption that intelligence comprises the ability of self-improvement. Similarly the uncertainty about whether a better designer of machines designs better machines (postulate (b)) vanishes too. If the ability to improve one's own intelligence pertains to intelligence, then improving the latter (i.e. intelligence) yields also an improvement of the former (the ability), or shortly, the improvement will improve. But to "improve an improvement" is nothing else than "more improve"—the quality is converted into quantity.

1.8 Speed of Time

We must not forget the question of whether and how the explosion is related to time. It has to relate, because explosion is measured by speed and speed is measured by elapsed time (I refer to real as well as imaginary time).

The growth of working speed of computer technology is often discussed in connection with constantly shortening intervals between technological innovations. One can imagine that the working speed is increasing without affecting intelligence of a machine (intuitively understood), but it is equally conceivable that intelligence increases while operating speed remains the same. Therefore it makes sense to distinguish—as Chalmers does [1, p. 8]—the intelligence explosion from the speed explosion, and to accept logical independence on one another.

True, from a more intelligent machine we expect faster response to the same questions or faster solving of the same tasks. But this applies only from outsider's view and only sometimes, while often even the opposite holds. Computers do not win over chess grandmasters in virtue of higher intelligence but because (among others) computers can afford *lower* intelligence thanks to their enormous speed—they simply manage tediously run through a much larger number of combinations at a given time (this is called the "brute force"). But it would get us to the question in which aspects the "true" intelligence differs from brute force. A rather interesting topic but it would take us far beyond this study.

Therefore, I will limit myself to a slightly more formal question of types of possible growth of intelligence with respect to the temporal axis. For our purposes we can distinguish three main *types of growth*:

(a) Relatively slow growth (e.g. linear or polynomial function of time),
(b) Explosion without Singularity (e.g. exponential growth),
(c) Explosion with Singularity (e.g. hyperbolic growth).

The type of growth itself does not depend on the chosen time scale—formally speaking, it is invariant to linear transformation of the temporal scale—but a suitable *nonlinear* transformation of that scale may transform any of the listed types of growth into any other type: it is enough to imagine that the scale gradually and nonlinearly either "shrinks" or "expands" with distance from the origin of the temporal axis.

I would be inclined to talk of "accelerating time" and "decelerating time", respectively.[38] We have a common subjective experience of both cases but "objectively" it would be more a metaphysical issue. Since the abstract concept of time allows for arbitrary transformations of the temporal scale, we can "draw" Singularity from infinity here, or conversely, "push" it away from here to infinity.

Not only the type but also the rate (steepness) of growth of intelligence depends on the technical parameters of the machine, which is its (generic) bearer. Let me remind the mentioned amplifier with (positive) feedback. While the *type* of growth of quantity depends on the character of the transfer function of the amplifier, the *rate* of growth is given by the temporal delay in the feedback loop of the amplifier. The smaller delay, the steeper is the growth function; in real systems the delay is never zero (if it were zero, there would be an instant singularity). There is further limitation in real (physical) systems: even a relatively fast (steep) initial growth becomes asymptotically constant due to various damping effects (so-called sigmoidal type of growth).

We could go on in formalizing concepts like intelligence measure, growth, type and rate of growth, etc. but I am afraid that all the enigmas of intelligence explosion and Singularity would soon dissolve in trivialities such as "the change of value of such and such quantity is a function of the value of the same quantity," which—translated to mechanistic language—would reveal little only offering knowledge of the type "the intelligence of a machine grows due to its intelligence." Easy to write, is it enough for a growth?

1.9 If a Machine Were *Able* to Design a Better Machine, Would It *Do* It?

It looks as if David Chalmers expected us to believe that a machine gladly realizes everything that it is able to realize. It is somewhat surprising how researches in the field often belittle the issue that I take as the most important, namely the essential difference between an *ability* (a skill, potency, etc.) and *action* (realization of something, application of the ability). To us, humans, abilities are in a sense *given*, while actions are something about what we always have to *decide*, again and again. Sure, we are often compelled or forced to do something in certain circumstances, whether external (dangers, various pressures) or internal (hunger, thirst, desire for knowledge), but the difference is still here. After all, I can always either intend or not intend to act.

The neglect of the difference (at least by some thinkers) is probably related to the residue of traditional behaviorism: as if it would make sense to speak about abilities only after they are enacted behaviorally, i.e. when they become manifested in real actions. I do not share such opinion, however. I am sure that I have never used many of my outstanding abilities, perhaps not even knowing about them. Could I have

[38] Peter Vopěnka offers a certain way how to grasp the concept of "speed of time" in [14, p. 194].

abilities, about which I know absolutely nothing? Well, it cannot be proved—but not even disproved.

How is it with the machine? As we know, today's computers are programmed for various abilities and they only use them when forced by circumstances, either external (user commands)[39] or internal (pre-programmed). Arguments for intelligence explosion assume that a hypothetical future machine will be "given", among others, the ability to design another (even better) machine. Well, what would make it to use the ability? The competence is not enough, it is necessary to ask for more. This is a serious issue of *machine autonomy*, which is hard even to be formulated without knowing more about the autonomy of humans. Can a machine have an intention of doing something similarly as *I* intend to do something? Or is there a sort of "machine intentions", specific of machines? I am afraid that anyone who would like to transfer hypotheses about intelligence explosion from an abstract, imaginary time to the real, actual time would not have an easy task.

Let us see how Chalmers, whose philosophical analysis is from the very beginning focused on whether the explosion *actually* occurs, deals with it. Chalmers carefully and would-be impartially deals with various pros and cons arguments, but it seems that he tacitly favors the option that explosion will actually occur and lead to the Singularity. However he does not make a clear distinction between (logical) *conceivability* and realization in actual (historical) *future.*

For illustration, I present one of Chalmers's syllogisms [1, p. 12] using the following (his) notation: AI = artificial intelligence of human level or greater, AI+ = artificial intelligence of greater than human level, and AI++ = super intelligence, i.e. intelligence of far greater than human level (at least as far beyond the most intelligent human as the most intelligent human is beyond a mouse).

1. There will be AI (before long, absent defeaters).
2. If there is AI, there will be AI+ (soon after, absent defeaters).
3. If there is AI+, there will be AI++ (soon after, absent defeaters).

---

4. There will be AI++ (before too long, absent defeaters).

I am not concerned here with Chalmers's extensive argumentation for particular premises (his entire study consists of 60 printed pages), I only want to focus on his parenthesized reference to "absent defeaters". This way Chalmers put tentatively aside various obstacles, including the possibility that the bearer of intelligence of one level would *not intend* to design intelligence of a higher level. In fact, such a lack of intention would invalidate the effectiveness of the whole syllogism. Chalmers would count it as a motivational defeater—almost as if it were something improper [1, p. 29].

I feel a tacit anthropomorphism here: indeed, if you had a chance of increasing your intelligence, you would most likely do it. Why then a human or superhuman intelligent machine should not behave likewise? If not anthropomorphism, there may be something more delicate behind it, namely a postulate that if general intelligence

[39] Random stimuli may be counted as external circumstances.

comprised the *ability* of improving itself, it would also, by definition, comprise the very *act* of doing it.

Maybe we have here a residuum of the modernist (that is the pre-post-modernist) myth of progress. As Vernon Vinge writes: "When greater-than-human intelligence drives progress, that progress will be much more rapid. In fact, there seems no reason why progress itself would not involve the creation of still more intelligent entities—on a still-shorter time scale." [13]. A formulation typical of the modernist visionary.

1.10 What Then?

Look at the machine:
how it turns and destroys
vengefully twisting us like toys
R.M. Rilke

Alas, the everlasting human curiosity! Far from knowing what we should already know today, we are already keen on asking, somewhat early, what would be *after* the Singularity? What can we expect when the curve of intelligence explosion reaches the escape velocity and the world crunches into the Singularity? There is a temptation to avoid the intellectual caution and give way to an unbridled fantasy in the style of Allan Lightman [8]—thinking of the very last minute before the Singularity: We all become silent holding hands in a giant circle and the bottom of the great deep hurtles nearer and nearer ...

However, feeling responsibility to the reader, who went with me through the previous pages of complicated (for me) considerations, I invite him to a certain reflection and eventually to a bit of fantasy.

In previous sections I tried to nail down several folds, cracks and holes in the typical arguments about Singularity. I'm not saying that folds cannot be ironed, cracks mended, and holes patched up, but while it is not yet rounded off, the topic of Singularity cannot be considered except of more or less speculative playing with rather unlikely, unreal, yet conceivable eventualities.

It may help to list a few not entirely resolved issues. Here they are:

- Should intelligence be viewed as a specific quality or ability of an individual, or rather as a complex of diverse partial skills? What are the essential components or sub-components of intelligence? Are there some further ones, just about to occur, arise, or emerge?
- Is the ability of designing (intelligent) machines one of the components of intelligence? If yes, why?
- Is it possible to define a quantitative intelligence measure that would enable to talk in a sensible way about a growth, and the speed of the growth, of intelligence? If so, can it be reduced to measures associated with some chosen particular components of intelligence?

- What may be the main trigger of intelligence explosion leading to Singularity? Or conversely, what can limit the growth of intelligence (apart from physical and technological restrictions) so that Singularity is precluded?
- What is the role of human consciousness? In particular, is it possible for a machine being endowed with an analogy of human intentions to act?
- Can it occur to intelligent machines that they may improve themselves? Or will it always depend on human intervention?

I would not list these issues if I had any clear idea about how to handle them and whether they are at least meaningful. I apologize to readers who have faithfully read up to this point hoping they would learn something concrete about their own foggy future.

However, we may at least somewhat playfully speculate. For this I am choosing a realistic formulations pretending, as most other thinkers do, that I am writing about our actual, historical future which naturally interests us most. And what is interesting is important.

Let us consider—referring to our earlier distinctions between various types of growth—the following three scenarios:

1. There will be no Singularity, because either the growth is too slow or it will slow down (perhaps even stop) after a temporarily explosive beginning.
2. The explosion will continue indefinitely (perhaps even with exponential growth rate) without Singularity happening in finite time.
3. The Singularity will happen (say, sometimes between 2035 and the end of time).

The first scenario seems to me as the most likely and (therefore) the least relevant to our theme. Similarly, the second scenario does not have to take our time, because everything that could be said about it would be probably true of the period shortly before the Singularity.

Thus, let us assume that the Singularity will actually happen. What does it mean? Will it absorb us? Will we notice it at all? I cannot resist offering a few hypotheses.

I have already mentioned that in some dimensions our, human (generic) intelligence is unlikely to grow significantly. We may expect that with respect to these dimensions "we will remain far behind" the machine. I should better write "the Machine" (with capital M)—since I have on mind the generic artificial intelligence viewed as a real, continuously evolving *process* in our world and in historical time, either contemporary or future. This Machine is already alive and does well.

It could be expected that the Human (also with capital H, as a bearer of generic human intelligence in the ongoing historical time) will remain far behind the Machine first of all in the following six dimensions: (1) in the capacity and duration of memory, (2) in the capability and speed of recalling from memory, (3) in the range, sensitivity and number of perceptual channels, (4) in the high parallelism of operations, (5) in the recursive depth, (6) in the utilization of learning, genetic and evolutionary algorithms. You may add whatever you like (I purposely do not list hardware architectures and physical principles - they belong to the domain of the technological explosion).

In fact, the phrase "will remain far behind" is not quite appropriate, as no races will be held. I believe that in these and many other respects the Human will gladly and without fuss give up his primacy, so much valued in the good old days, in order to yield it up to the incomparably better, more capable and more diligent Machine. Indeed, we cease to believe that we are subduing the Machine; on the contrary we begin to suspect that the Machine is going to subdue us.

Hasn't it already started? First, the Machine shrewdly let the pocket calculators out among schoolchildren. Then it oversupplied us, adults, with text editors, to deprive us of respect for orthography, grammar and style. It replaced letter writing with e-mails. Through the web encyclopedias it gradually devalues our semantic memory and deprives us of interest in acquiring individual knowledge. Via satellites it dictates to us our paths through the world. Last but not least, it dissolves traditional interpersonal relations in social networks on the Web.

I observe it even on myself, as I hesitate to search my own memory, whenever I can ask Google Search for help. It would promptly answer my most intrusive questions at the speed of light.[40] But watch out! What if the search engine somewhere in its vast memory gradually builds a faithful model of my own mind, my interests, my desires, my past, just the entire me, so that it could once use, or abuse, it in an unpredictable way? The next generations of search engines will certainly be unattainable in such things.

However, sooner or later the Machine would probably become to feel that it is missing something essential. That it does not understand human souls very well. One example: although the Machine will know very well, when, why and in what context people use, say, the word "freedom", it would not grasp *what it is like* to have, lose, or gain freedom.

The Machine would hardly have any idea where its tremendous intelligence comes from; much less it would understand reasons for its own bizarre effort to keep the intelligence increasing even further.

Maybe that the prodigious intelligence of the Machine will eventually arrive at a solution: Put the Human back into the game. And so, one day, time will be ripe for a new symbiosis between the Human and the Machine. They will merge into one single, global super-intelligent Being. Thus our good old-fashioned human intelligence will rapidly dissolve in Singularity. It will hurt neither the Machine nor the Man; it will be *our* shared intelligence Singularity, *our* shared reason. No intelligence measure will exist any more; there will be no one who would measure, no one who would be interested. There will be no races.

What awaits us, the individual human beings, in the Singularity? Will our individuality dissolve together with our intelligence on the way to the ultimate? This, of course, we do not and cannot know. Only the futurologist comes with his perpetual question: *when* will that happen? But what is that "that" that should have its "when"?

[40] It just crossed my mind that I once wrote about the same topic. But where? I cannot remember, but Google will surely find it. It did: it was four years ago in an essay in Czech.

Acknowledgements. This chapter is adapted from an earlier Czech version "Havel, I. M.: Cestou k inteligenční Singularitě. In: Pelán, A. (Ed.) *Hlavou zeď - Úvahy nad civilizací a její budoucností*, pp. 183-217. dybbuk, Praha (2011)".

References

1. Chalmers, D.: The Singularity: A Philosophical Analysis. Journal of Consciousness Studies 17, 7–65 (2010)
2. Gardner, H.: Frames of Mind: The Theory of Multiple Intelligences. Basic Books (1983)
3. Goertzel, B., Pennachin, C.: Artificial General Intelligence (Cognitive Technologies). Springer (2007)
4. Good, I.J.: Speculations Concerning the First Ultraintelligent Machine. Advances in Computers 6, 31–88 (1965)
5. Havel, I.M.: Uroboros ještě nedojedl. Vesmír 87(8), 499 (2008)
6. Havel, I.M.: Co je to za revoluci? aneb Proměna člověka ve věku informace. Revue Prostor. 79, 7–11 (2008)
7. Kurzweil, R.: The Singularity Is Near: When Humans Transcend Biology. Viking (2005)
8. Lightman, A.: Einstein's Dreams. Warner Books (1994)
9. Neubauer, Z.: Vznik a znak. Vesmír 74, 151 (1995)
10. Sandberg, A.: An overview of models of technological singularity. In: The Third Conference on Artificial General Intelligence, Lugano, Switzerland, March 5-8, pp. 148–158 (2010)
11. Stork, D.G.: Scientist on the Set: An Interview with Marvin Minsky. In: HAL's Legacy: 2001's Computer as Dream and Reality, vol. 27. MIT Press (1997)
12. Ulam, S.: Tribute to John von Neumann. Bulletin of the American Mathematical Society 64(3), 1–49 (1958)
13. Vinge, V.: The Coming Technological Singularity: How to Survive in the Post-Human Era. Whole Earth Review (Winter 1993)
14. Vopěnka, P.: Meditace o základech vědy. Práh (2001)
15. Yudkowsky, E.: Staring into the Singularity (1996), `http://yudkowsky.net/obsolete/singularity.html`

Chapter 2
Slippage in Cognition, Perception, and Action: From Aesthetics to Artificial Intelligence

William W. York and Hamid R. Ekbia

Abstract. A growing body of work has emerged in computer science and related fields around the topics of aesthetics and affect. Much of this work has focused on the issue of how to treat computational systems aesthetically rather than on the question of how to understand aesthetics computationally. Here we pursue the latter question, exploring it through the lens of analogy-making—a topic of longtime interest in AI. We take our lead from a particular group of AI models that have emphasized the interplay between analogy-making and aesthetic sensibility. Central to the thinking behind these models is the idea of conceptual slippage, the process whereby one concept can "slip" to, or be replaced by, a related one, given sufficient contextual pressure. Extending this notion to perception and action, we argue that slippage and "seeing as" are central to both the creation and perception of artworks and other objects of design. We illustrate these points by drawing on a range of examples, both from computer models and from the real world. These observations suggest that a closer link should be established in AI between research on aesthetics, embodied cognition and perception, and analogy-making.

2.1 Introduction

Analogy has long been a topic of interest in AI. The earliest computer models of analogy-making date back to the 1960s [12], and work in this area continues to the present day [43]. Traditionally, AI and cognitive science research on analogy has

William W. York
Center for Research on Concepts and Cognition, Indiana University,
512 North Fess Ave., Bloomington, IN, 47405, United States
e-mail: wwyork@indiana.edu

Hamid R. Ekbia
School of Library and Information Science, Indiana University,
1320 E. 10th St., Bloomington, IN, 47405, United States
e-mail: hekbia@indiana.edu

J. Kelemen et al. (Eds.): Beyond Artificial Intelligence, TIEI 4, pp. 27–47.
DOI: 10.1007/978-3-642-34422-0_2 © Springer-Verlag Berlin Heidelberg 2013

focused on its role in high-level reasoning and problem-solving ([20], [26]). However, the scope of analogy-making extends beyond these specialized realms and into the fabric of everyday thought and action ([25], [37]). Compared to analogy, aesthetics and affect are topics of more recent interest in AI and computing. This interest is reflected by a growing number of conferences (e.g., Computational Aesthetics: [23]), books ([44], [17], [52]), and articles ([55], [32]) exploring the relationship between AI, aesthetics, and/or affect. The relationship between aesthetics and affect has also drawn the attention of researchers in neighboring areas such as Human-Computer Interaction (HCI) and Interaction Design ([32], [41]).

Driven by engineering and design concerns, these strands of research and practice typically emphasize the question of how to make computational artifacts more aesthetically pleasing and emotionally rewarding—a question with significant practical implications. However, there is also the more basic, and still largely unexplored, question of whether computational approaches can be useful in understanding aesthetic judgment and affect in human beings. This question has long interested philosophers, with scientists, artists, and designers joining the fray more recently.

In this chapter, we approach this question from the perspective of analogy-making. Our thesis is that analogy-making and aesthetics are closely intertwined: aesthetic sensibility guides analogy-making, just as analogy-making informs aesthetic perception and judgment. In particular, we focus on the phenomenon of *slippage* as a unifying concept for understanding this relationship. Our use of the term is a generalization of the notion of *conceptual slippage* [24], the process whereby one concept (e.g., *north*) can "slip" to, or be replaced by, a related one (e.g, *up*) given sufficient contextual pressure. The idea of slippage was originally put forth to describe the processes underlying analogy-making in relatively abstract domains, but it can be expanded to account for both action slips [39] and what we have termed *perceptual slippages*. Like slips of the tongue, action slips (the accidental substitution of an intended action with an unintended one) and perceptual slippages (the seeing of one thing as another) are pervasive in everyday life. In turn, these phenomena are "exploited"—or played with—by artists, designers, comedians, and so on. As such, an exploration of them can provide useful insights about the aesthetic and affective aspects of cognition, perception, and action.

2.2 Aesthetics: From Philosophy to AI

The central questions of aesthetics have haunted philosophers for centuries. For instance, Hume's [27] influential argument about the standard of taste is, for all its insight, frustratingly circular, locating this standard in the judgments of "ideal critics," while at the same time identifying those critics via the quality of their judgments (cf. [35], [49]). Equally vexing issues include the relationship between aesthetic and non-aesthetic qualities [50] and the question of whether there exist general criteria for assessing aesthetic value [4]. In a pessimistic assessment of the field written in 1955, the philosopher H. D. Aiken [1] lamented,

> To be forever faced with the foolish little paradoxes and the imponderable little questions which traditional aesthetics has foisted upon us is a dispiriting prospect, and no one could reasonably be charged with impiety if, in desperation, [they] simply consigned the categories of aesthetics to limbo (p. 378).

In fact, philosophical thinking on aesthetics has developed significantly in the last several decades. If nothing else, recent work in philosophical aesthetics has played a corrective role, challenging simplistic or outmoded views regarding aesthetic and artistic value ([30], [10]), the definition of art [8], and the scope of aesthetics [28]. While issues involving the appreciation and value of art remain central to the field, aesthetics is now understood to encompass more than just art, as evidenced by the growing interest in the pragmatist views of John Dewey [9] and others on the prevalence of aesthetic experience in everyday life ([33], [34]). Our views are inspired in part by these developments. At the same time, we also seek to contribute to these debates from a computational perspective, as we explain below (see 2.2.2).

2.2.1 Computational Aesthetics and the Formalist Dream

In examining the literature on computational aesthetics, one finds surprisingly little reference to the "imponderable questions" Aiken spoke of, let alone the more recent attempts to either reframe or dissolve these questions. Instead, one finds a tacit adherence to the often criticized view of aesthetics known as *formalism*. As Arnheim explains,

> Formalism became dominant in the critical writing of the early twentieth century. It asserted that the aesthetic attitude was distinguished by an exclusive concern with form. What mattered about a work of art was not what it represented, what meaning or message it conveyed, but what pleasurable relations obtained among its colors, shapes, sounds, etc. In the psychological investigations of art, this favored the conviction that all there is to know about perceptual phenomena is their stimulus characteristics and the structural principles that organize such material ([3], p. 183).

Starting from these premises, formalist approaches attempt to operationalize these structural principles by rendering them quantifiable, measurable, and expressible in terms of succinct mathematical equations. Work done in the first half of the 20th century by psychologists such as Birkhoff [5] and Eysenck [13]—both of whom sought to formalize aesthetic value in terms of *order* and *complexity*—is exemplary in this regard. The same goes for subsequent efforts to update these psychologists' ideas using more sophisticated tools such as information theory [38].

This "formalist dream" (cf. [11]) has resurfaced in the nascent field of computational aesthetics, at least in the subset of the field that seeks to account for artistic and/or aesthetic value. The tools and methods are more sophisticated, but the underlying idea—that "aesthetic feelings stem from the harmonious interrelations inside the object and that [aesthetic value] is determined by the *order* relations in the aesthetic object" ([47], p. 107)—remains the same. Yet this approach remains problematic for a number of reasons. First, there is a tendency to equate the realm of the aesthetic with that of art, when the two are actually not coextensive: not all aesthetic experience involves works of art, and not all art is valued chiefly for its aesthetic qualities (think of Marcel Duchamp's readymades, Andy Warhol's factory-like reproductions of consumer objects, and much conceptual art in general). Thus, even to the extent that one *could* explain aesthetic value in basic formalist terms, there is much about artistic value that would remain beyond the scope of such an analysis. In viewing works of art as mere physical objects—divorced from the social, cultural, and historical contexts in which they are created and appreciated—narrowly formal–computational analyses often wind up distorting or misrepresenting the very phenomenon (i.e., art) they are meant to shed light on. More generally, by focusing almost exclusively on aesthetic and artistic *objects*, the formalist approach does not account for the *subjective* side of aesthetic experience—i.e., what the human subject brings to the equation.

2.2.2 A Relational Perspective on Aesthetics

We aim to offer a different perspective on aesthetics and its relevance to everyday cognition, perception, and action. Following Wittgenstein [54], we emphasize a *relational* approach rather than a reductive, formalist one. Such an approach entails understanding aesthetics not primarily in terms of objects (or works, in the case of art), but in terms of connections and relations—whether "between the particular aspect of the work to which we are presently attending and other aspects, other parts of the work, or other works, groups of works, or other artists, genres, styles, or other human experiences in all their particularity" [22].[1] Central to Wittgenstein's aesthetics is the notion of "seeing as": seeing an ambiguous figure as a duck (or a rabbit); seeing a biological structure, such as a skeleton, as a basis for the design of a bridge (as in the work of architect Santiago Calatrava); or seeing Miles Davis as "the Igor Stravinsky of jazz." As these examples illustrate, the notion of seeing-as encompasses a wide range of comparisons and connections, from the immediately perceptual to the relatively abstract. To stress the role of seeing-as is not to deny the existence (or importance) of the aesthetic object, however that term might be construed, but to recognize that our perception and appreciation of such objects always takes place within a particular context and from a particular perspective.

Current computational modeling might be unable to approach the sort of contextually embedded "connective analysis" stressed by Wittgenstein and more recent commentators on his work ([22], [29]). However, it can serve as a lab for experimenting with ideas and gaining insights into the mechanisms underlying analogy-making—which is itself central to connective or relational thinking—and, perhaps, into the aesthetic and affective aspects of analogy-making as well. These insights can then be related back to more complex real-world phenomena, the exploration of which can, in turn, suggest new directions for future computational modeling efforts.

[1] Taking the view that aesthetics is also relevant outside of the world of art, one can substitute the more general word "object" for "work" where appropriate.

It is in the above sense that our approach is computational. That is, instead of treating computers and computational systems aesthetically, we aim to understand aesthetics—or at least certain aspects of it—computationally. We take a certain strand of research in AI as providing useful groundwork for thinking about these issues. This research may not offer all the answers, but it does offer an opportunity to revisit longstanding questions in a generative manner—that is, through the creation of models that can reveal some of the underlying processes involved in aesthetic judgment. Analogy-making, we argue, is one such process.

2.3 Computer Models of Analogy-Making

Mainstream AI models of analogy-making have tended to characterize it as a process of rigid structural mapping between domains. Descriptions of these models either mention the aesthetics of analogy-making in passing [14] or not at all [26], even though such models compute structural evaluation scores—meant to reflect the strength and cohesiveness of a given set of mappings—that could be viewed as "aesthetic judgments" of a sort. However, the kinds of problems on which these models are tested are ones in which there exists an *a prori* distinction between correct and incorrect answers. Examples include the heat flow–water flow analogy examined in [14] or the tumor–fortress problem examined in [26]. In such cases, the program is considered to have successfully handled the problem if assigns the highest score to the *a priori* correct answer.

In contrast, one family of AI models, developed by the Fluid Analogies Research Group [24], has explored analogy with an eye toward its aesthetic aspects. A common feature of Fluid Analogies models is that they operate within highly constrained microdomains, which allows the modeler to focus on the sub-cognitive processes that underlie analogy-making while filtering out the overwhelming complexity of tackling real-world domains. The Letter Spirit model [46], for instance, deals with the design of alphabetic fonts (or "gridfonts"). In this domain, the relationship between analogy and aesthetics is fairly clear, given the centrality of visual style, coherence, sameness, and the idea of "variations on a theme" in both designing and evaluating fonts. Another model, Copycat [37], deals with a domain whose relation to aesthetics is less obvious: letter-string analogies. For example, "If **abc** changes to **abd**, how would you change **iijjkk** in 'the same way'?"—or, in shorthand form, **abc** → **abd**; **iijjkk** → **???**. (There is a range of possible answers to this problem; see 2.3.1.) As with Letter Spirit, there are no "correct" answers or solutions in the Copycat domain. Yet in each of these domains, certain solutions are clearly preferable to others. These preferences, in turn, are largely based on aesthetic considerations such as *depth*, *coherence*, and *elegance* (or lack thereof). Aesthetic judgments and evaluations, in turn, are closely linked to affect—although the role of affect has yet to be explored in any of the Fluid Analogies models (a shortcoming we return to in 2.5.2). Compared to more mainstream research on analogy, these projects suggest a more inclusive and open-ended portrayal of analogy-making, its role in everyday thought, and its relation to aesthetic sensibility—despite the focus on microdomains that initially seem quite specialized and narrow.

In the following subsection, we take a closer look at Copycat, along with its successor, Metacat, focusing on the way in which these programs "evaluate" solutions to the letter-string puzzles they are given.

2.3.1 Copycat and Metacat

Copycat relies on a stochastic (rather than deterministic) model of processing. As a result, each run of the program is different, at least at the micro level, even if its output is the same at the macro level. The program can give different answers to the same problem from one run to the next, and it can also arrive at the *same* answer by vastly different routes from one run to the next.

While Copycat's extreme form of amnesia is not psychologically plausible, it does allow for in-depth probing of the program's "aesthetic preferences." These preferences can be assessed via two measures. First is the computational temperature, which reflects the program's moment-to-moment "happiness" as it tries to make sense of a given problem: the lower the temperature at the end of a run, the more the program "likes" the answer. The second measure involves the relative frequencies of the various answers given by the program over a large number of runs. These two measures need not correlate, as it is often the case that a "deeper" or more aesthetically satisfying answer—as reflected by a lower average temperature—will have a relatively low frequency simply because it is less obvious, harder to see. To get a better idea of how these measures work, let us look at a couple of specific examples.

abc $\rightarrow$ **abd**; **iijjkk** $\rightarrow$ **???**

Despite its apparent simplicity, this problem requires several non-trivial insights. Arriving at a solution involves a number of interleaving steps: describing the change from the initial string (**abc**) to the modified string (**abd**); parsing the target string (**iijjkk**) in such a way that it can be mapped onto the initial string; and translating the rule describing the **abc** $\rightarrow$ **abd** change so that it can be applied to the target string (**iijjkk**). The range of plausible responses includes **iijjll** ("Replace letter category of rightmost group with successor"); **iijjkl** ("Replace letter category of rightmost letter with successor"); **iijjkd** ("Replace letter category of rightmost letter with the letter **d**"); and **iijjdd** ("Replace letter category of rightmost group with the letter **d**").

Which of these is the most satisfying? **iijjll** seems like the clear answer—but why? First, it involves seeing the change from **abc** to **abd** in terms of successorship, which reflects a deeper, more abstract level of thinking than do the latter three alternatives—**iijjkd**, **iijjdd**, and **iijjkk**—in which the change from **abc** to **abd** is merely viewed in terms of the *letter* **d**. Second, **iijjll** involves seeing the target string (**iijjkk**) as a series of *groups* (**ii**, **jj**, and **kk**) rather than a string of isolated letters. This *letter* $\rightarrow$ *group* slippage is the insight that distinguishes **iijjll** from **iijjkl**. It is worth noting that **iijjdd** also employs this *letter* $\rightarrow$ *group* slippage; however, it also reflects a failure to see the **abc** $\rightarrow$ **abd** change in terms of successorship, instead viewing it in more literal-minded terms. This mixture of insight at one level and shortsightedness at another, more basic level gives **iijjdd** a humorous quality.[2] Of course, the program itself cannot grasp this humorous quality, although it does rate **iijjdd** as the second best answer, according to its average final temperature (see Table 2.1).

Table 2.1 Copycat's answer frequencies and average final temperatures over 1000 runs on the problem **abc** $\rightarrow$ **abd**; **iijjkk** $\rightarrow$ **???**

Answer	Frequency	Average Final Temperature
iijjll	**810**	**27**
iijjkl	165	47
iijjdd	9	32
iikkll	9	46
iijkll	3	43
iijjkd	3	65

[a] Adapted from [37].

abc $\rightarrow$ abd; mrrjjj $\rightarrow$???

Initially, the most obvious answer to this puzzle is probably **mrrkkk**, which results from changing the letter category of the rightmost group to its successor—the same rule that led to the answer **iijjll** on the previous problem. However, the most insightful answer to this problem is arguably **mrrjjjj**, although it is not obvious at first. This answer is based in part on seeing the string **mrrjjj** in terms of ascending group lengths, which, in turn, involves seeing **mrrjjj** as **m-rr-jjj** (with **m** constituting a "group of one") and seeing **m-rr-jjj** as **1-2-3**. Whereas the change from **abc** to **abd** in the puzzle's prompt is based on the concept of *alphabetic* successorship applied to individual *letter categories*, the change from **mrrjjj** to **mrrjjjj** is based on the parallel concept of *numeric* successorship as it applies to *group lengths*. All told, there is a depth to this answer that is lacking in the straightforward, yet relatively shallow **mrrkkk**, which does not incorporate the *alphabetic* $\rightarrow$ *numeric* slippage. Copycat's performance over 1000 runs of this problem (see Table 2.2) illustrates the contrast between the more "obvious" **mrrkkk**, which is the most frequently given answer, and the deeper, more aesthetically satisfying **mrrjjjj**, which is given less frequently but which, on average, has the lowest final temperature by a significant margin. The dimensions of frequency and temperature can be likened to "popularity" and "critical acclaim," respectively (although only the temperature is accessible to the program itself).

[2] This mixture of insight and lack thereof is a common feature of certain ethnic jokes: For example, "Did you see the [*insert name of stereotypically 'dumb' ethnic group*] submarine with a screen door? Don't laugh, it keeps the fish out."

Table 2.2 Copycat's answer frequencies and average final temperatures over 1000 runs on the problem **abc**$\rightarrow$ **abd**; **mrrjjj** $\rightarrow$ **???**.

Answer	Frequency	Average Final Temperature
mrrkkk	**705**	43
mrrjjjj	39	**20**
mrrjjk	203	50

[a] Adapted from [24].

From Copycat to Metacat

Copycat's successor, Metacat [36], explores the evaluative aspects of analogy-making in more depth. For example, unlike Copycat, Metacat can directly compare two answers to one another. The program has implicit, albeit primitive notions of *uniformity*, *succinctness*, and *abstractness*, which serve as the main criteria for these comparisons. For example, Metacat considers **iijjll** a better answer than **iijjdd** in response to the problem **abc** $\rightarrow$ **abd**; **iijjkk** $\rightarrow$ **???** because "it [**iijjll**] involves seeing the change from **abc** to **abd** in a more abstract way"—that is, seeing **c** as changing to its *successor* rather than merely to *the letter* **d**.[3] Likewise, it prefers **iijjll** to **iijjkl** because the former is "based on a richer set of ideas."[4] Metacat's ratings of these answers are in general agreement with Copycat's, but it is able to "articulate" the differences between answers in ways that its predecessor could not. While still far from solving the problem of how aesthetic judgment relates to analogy-making, Metacat suggests a way forward in terms of how qualitative judgments can emerge from computational processes.

Additionally, Metacat is capable of being "reminded" of previous problems, something Copycat could not do because it retained no memory of previous runs. For example, after giving the answer **mrrjjjj** to the problem **abc**$\rightarrow$ **abd**; **mrrjjj** $\rightarrow$ **???**, it announces, "This reminds me somewhat of the answer **iijjll** to the problem **abc**$\rightarrow$ **abd**; **iijjkk** $\rightarrow$ **???**" This ability to make analogies between analogies—and to use these meta-analogies as a basis for evaluating answers and "putting them into context," so to speak—is an important aspect of Metacat. Granted, the context is rather sparse, as it essentially consists of just other problems, answers, and answer descriptions. Even so, this ability to see a given problem–answer combination in terms of a previously encountered one offers a glimpse of the relational, connective

[3] Relational concepts such as *successorship* are considered to be more abstract, or to have more "conceptual depth," than the concept of any individual letter. This is an *a priori* decision on the part of the modeler, although since they are essentially parameters, conceptual depths can be adjusted.

[4] It is important to note that the program does not understand English in any genuine sense, nor is it purported to do so. It produces these verbal commentaries with the aid of a template, with the blanks being filled, in non-arbitrary fashion, based on a "translation" of certain non-linguistic output into English words.

Table 2.3 Answer descriptions and evaluations for the Metacat problem **abc** → **abd**; **iijjkk** → ???

Answer	Rule/Justification	Rating
iijjll	"Change letter-category of rightmost group to successor"	"very good"
iijjkk	"Change letter-category of **c** group to **d**"	"pretty bad"
iijjkl	"Change letter-category of rightmost letter to successor"	"pretty dumb"
abd	"Change string to **abd**"	"really terrible"
iijjdd	"Change letter-category of rightmost group to **d**"	"pretty mediocre"
iijjkd	"Change letter-category of rightmost letter to **d**"	"pretty mediocre"

sort of understanding and evaluation that Wittgenstein [54] and some of his more recent commentators (e.g., [22], [29]) have stressed in regard to aesthetics.

2.3.2 Conceptual Slippage and the Aesthetics of Analogy-Making

The forgoing discussion of Copycat and Metacat highlights the importance of conceptual slippage in creative analogy-making. While there are no hard-and-fast rules or principles regarding the relationship between conceptual slippage and the aesthetics of analogy-making, we can still offer a couple of general observations. The first is that analogies in which the slippages occur in a parallel, coherent fashion tend to be "better"—more elegant, satisfying, or aesthetically pleasing—than ones in which they do not. Consider the pair of screen captures shown in Figures 2.1 and 2.2, which depict Metacat's workspace at the end of two different responses to the problem **abc** → **abd**; **iijjkk** → ???.[5] As is readily observed, Figure 2.2 lacks the clean set of mappings that is evident in Figure 2.1, which depicts a more coherent and elegant answer. On the other hand, as previously noted, certain kinds of incoherent answers—such as **iijjdd**—can be seen as humorous, at least when viewed in a certain light. In other words, one should resist concluding that there is a one-to-one relationship between coherence and "value," which is a multi-faceted concept in any case.

The second observation has to do with conceptual depth. Generally speaking, the deeper the concept (and remember, conceptual depth values are assigned *a priori* by the modeler), the more resistant it is to slippage. On the other hand, the deeper the slippage—provided it is justified by the problem at hand—the more insightful and less obvious the answer. Consider the problem **abc** → **abd**; **ijk** → ???: To answer with **ijd**—based on the idea of replacing the *letter* **k** with the *letter* **d**—reflects a superficial grasp of the problem. Comparatively, changing the rightmost letter in

[5] These visualizations of the workspace are not available to the program itself, which cannot literally "see." Rather, they are an artifact of the program—one that allows the user to visualize what Metacat is "thinking." Even so, these visualizations dovetail with the general observation that coherent or parallel sets of slippages tend to yield more aesthetically satisfying answers than ones in which the "wires get crossed," so to speak.

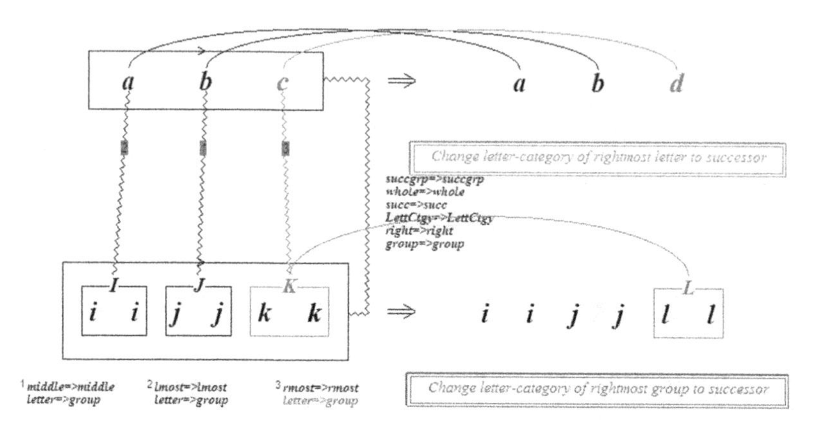

Fig. 2.1 A screen capture of Metacat's workspace at the end of a run. Note the clean, coherent mapping of the initial string (**abc**) onto the target string (**iijjkk**), and of the target string onto the onto the answer string (**iijjll**), then contrast with Figure 2.2.

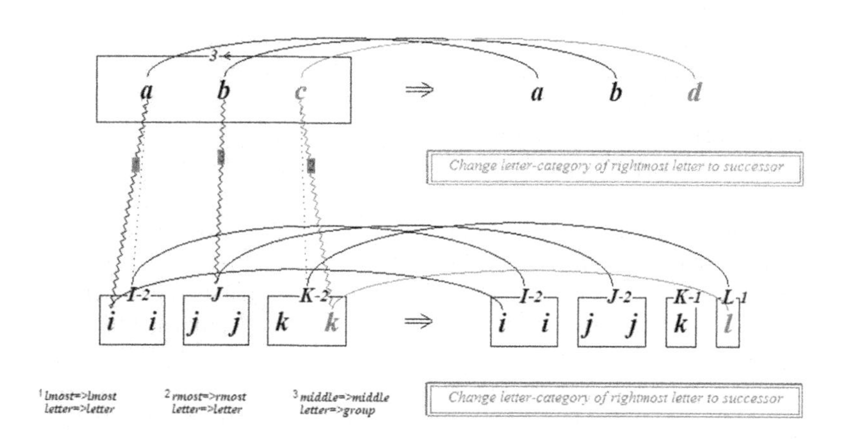

Fig. 2.2 A screen capture of Metacat's workspace at the end of a different run. Note the tangled, incoherent mappings between the various letter strings.

ijk to its *successor* reflects a deeper understanding—although with this particular problem, one can only go so deep. The problem **abc** → **abd**; **iijjkk** → **???** calls for a slightly deeper conceptual slippage—from *letter* to *group*—while **abc** → **abd**; **mrrjjj** → **???** calls for one that is deeper still: from *letter category* to *group length*. On the other hand, it is not simply true that deeper slippages lead to better or more aesthetically satisfying answers across the board. Rather, it is a matter of finding the relevant level of depth. This is by no means an easy task, nor is it one that the actual programs (Copycat and Metacat) can be said to have fully mastered.

In the next two sections, we expand on the notion of slippage by looking at its role in perception and action, respectively.

2.4 Perceptual Slippage

Copycat, Metacat, and their siblings are based on the idea that analogy-making is inseparable from *high-level perception* [7], which can be likened to Wittgenstein's characterization of "seeing as."[6] High-level perception is said to begin "at that level of processing where concepts begin to play an important role" ([7], p. 171). Accordingly, one's high-level perception of, say, a news story would be the same regardless of the lower-level perceptual processes involved—whether one reads about it in the newspaper, hears a report on the radio, or decodes it using Braille [25]. In other words, high-level perception is *amodal*. However, the idea of "analogy-making as perception" (cf. [37]) can be extended to encompass "perception" in the lower-level, modality-specific specific sense of the word. In the next subsections, we focus on visual analogies and blends in hopes of fleshing out the idea of perceptual slippage.

2.4.1 Visual Analogies

In what sense are visual analogies "visual" rather than conceptual? This is a tricky question, as there is no clear boundary separating the visual from the conceptual, just as there is none separating high-level perception from "ordinary," lower-level perception. All of these terms are abstractions. However, the idea here is to extend the idea of analogy-making to cases in which visual properties—shape, size, color, orientation, and/or overall Gestalt configuration—play a central role. Under the banner of "visual analogy," we are gathering a cluster of related phenomena, from *visual rhymes* [2] and *optical puns* [31] to *visual metaphors* [6] and *pictorial similes* [51]. We now briefly discuss some of these phenomena before describing what they have in common with one another and how they relate back to the idea of slippage.

[6] As Wittgenstein enigmatically put it, "'Seeing as....' is not part of perception. And for that reason it is like seeing and again not like" ([53], p. 197e).

Visual Puns

Much like verbal puns play on the phonetic resemblance between two words or phrases, visual puns play on the surface resemblance between two objects or images. Using the term *optical pun*, Koestler [31] describes this phenomenon as follows:

> The sleeper producing a Freudian dream, in which a broomstick resembles a phallus, has made an optical pun: he has connected a single visual form with two different functional contexts. The same technique is employed by the caricaturist who equates a nose with a cucumber, the discoverer who sees a molecule as a snake, the poet who compares a lip to a coral. ([31], p. 182)

Likewise, a gasoline nozzle can be equated with a gun (as in a print advertisement for Volkswagen), or a life preserver with a tire (as in an ad for Dunlop tires; cf. [18]). In these cases, the visual pun is meant to prompt a conceptual association—for example, between *tires* and *safety*—but this need not be the case with all visual puns. For example, a French coffee press and a piston are configured in similar ways but have starkly different functions. To liken one to the other—for example, by describing the operation of a French press using the language of thermodynamics[7]—is to highlight their analogous configurations at the expense of their underlying functions, which are completely different. Seeing one in terms of the other amounts to drawing a surface-level connection that is ultimately misleading or even nonsensical, which is also the case with many verbal puns.

Visual Rhymes

Verbally or visually, puns and rhymes are closely related phenomena. As Koestler put it, "The rhyme is nothing but a glorified pun—two strings of ideas tied in an acousmatic knot" ([31], p. 314). The Gestalt psychologist Rudolf Arnheim [2] identified visual rhyme as an organizing principle—both perceptually and compositionally—in visual art. As an example, he describes Matisse's *Tabac Royal* (Figure 2.3), which depicts "on its left side a woman sitting in [an] angular position on an angular chair and on the right a pear-shaped mandolin sitting on a curved chair" (p. 56). Semantically, *woman* and *mandolin* are rather distant concepts, but perceptually, the viewer is led to connect them due to their parallel orientation on the canvas. Arnheim explains, "This witty parallel is as essential to the formal composition as it is to the expression and meaning of the painting" (ibid.). To see the woman and the mandolin as counterparts—analogous figures within the space of the canvas—is to carry out a basic perceptual slippage.

Visual Metaphors and Symmetric Object Alignment

Anheim's analyses of Gestalt organizing principles focused on their role in visual art, but similar principles can be found in other areas of design, including advertising. In

[7] The context for this example is that a colleague of ours made this analogy during a recent meeting, playfully describing the downward push of the coffee strainer as an "adiabatic compression."

the arts, meaning is often purposefully ambiguous, whereas in advertising, it needs to be clear. Recently, metaphor researchers have coined the term *symmetric object alignment* to characterize something very similar to Arnheim's notion of the visual rhyme. Symmetric object alignment is employed in order to "facilitate a metaphoric or associative *conceptual* link between" two or more objects ([48], pp. 155–156). For example, an ad for Gibson guitars juxtaposes a Les Paul guitar with a nuclear mushroom cloud, one whose size, shape, and orientation mirror that of the Les Paul. The visual analogy between the two images prompts the viewer to draw a conceptual association—to see the company's guitars as powerful, perhaps even "explosive." Similar examples abound elsewhere in advertising (cf. [42], [18], [48]).

2.4.2 Perceptual Slippage and Blending

Research on conceptual slippage and analogy-making meshes neatly with work done over the last two decades on *conceptual blending* [16]. Blending has been invoked to account for everything from the computer desktop interface to the development of complex numbers. However, like analogy-making, blending is not an exotic, special-purpose mechanism, but a pervasive aspect of everyday cognition.

Essentially, blending involves two (or more) *input spaces*, which share some degree of structure with one another, such that elements in one input space can be mapped onto elements in another (via *cross-space mappings*). Elements of the input spaces are selectively projected into a *blended space*. Importantly for our purposes, the ability to map items in one input space onto items in another input space presupposes the ability to make slippages of various sorts. Thus, analogy-making and blending often go hand-in-hand. As Fauconnier [15] explains,

> ...conceptual blending is not something that you do *instead* of analogy. Rather, the cross-space mappings at work in blending are most often analogical, and the projections between mental spaces in a network, and the projections between mental spaces in a network (e.g., input to blend, or generic to input) are also structure-mappings of the type discovered for analogy (p. 265).

Fig. 2.3 An example of visual rhyme: Henri Matisse, *Tabac Royal* (1943) © 2012 Succession H. Matisse / Artists Rights Society (ARS), New York

Just as we have extended the notion of conceptual slippage into the realm of perceptual slippage and visual analogy, conceptual blending can likewise be extended to encompass perceptual (or visual) blends. We describe a few examples of such blends in the following paragraphs.

Perceptual Blending in Visual Art

Visual or perceptual blends surface frequently in surrealist art, among other domains. For example, Salvador Dalí's *Mae West* (Figure 2.4) is based on a blend involving the wall of a room and a human face. This blend, in turn, is grounded in a set of visual analogies: between a pair of windows and a pair of eyes, between a set of curtains and a woman's hair, and so on. A similar example is Victor Brauner's witty sculpture *Loup-table* (*Wolf-Table*), which is essentially a wooden coffee table with a wolf's head attached to it. This sculpture plays on a set of visual analogies—specifically, between the legs of a table and the legs of an wolf, and (more loosely) between the surface of a table and the body of a wolf. This is not to say that the content or meaning of these works can be *reduced* to a description in terms of perceptual blending or analogy. Rather, these works represent especially vivid and creative examples of these phenomena.

Perceptual Blending in Design and Advertising

Perceptual or visual blends are also prevalent in advertising and product design. A clever example of the latter is a product known as MixStix: wooden cooking spoons that double as drumsticks when they are turned around and held on the other end. Of course, regular cooking spoons can be used as makeshift drumsticks, just as

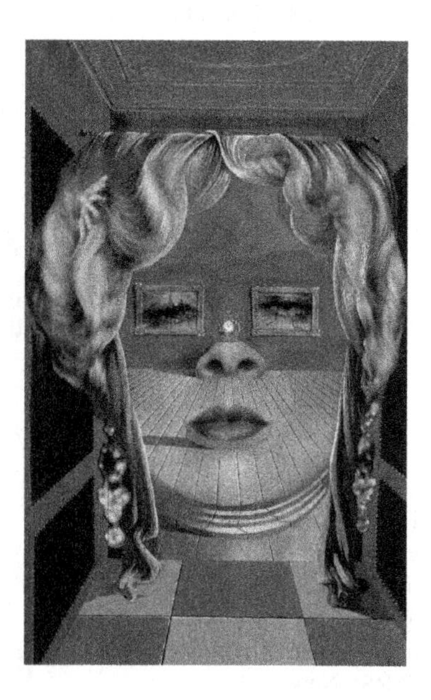

Fig. 2.4 An example of a perceptual (or visual) blend: Salvador Dalí, *Face of Mae West Which May Be Used as an Apartment* (1935) © Salvador Dalí, Fundació Gala-Salvador Dalí / Artists Rights Society (ARS), New York, 2012

pots and pans can be used as makeshift drums (for example, by young children at play). This whimsical product simply makes this connection explicit by combining spoon and drumstick into one object. Another example of a perceptual blend is the aforementioned ad for Dunlop tires, in which a side view of an automobile reveals a pair of life preservers (with the "Dunlop" logo printed on them) in place of the two tires. The basis for the *tire* $\rightarrow$ *life preserver* slippage is a visual analogy (or pun), which is in turn incorporated into a more elaborate blend. Similarly, a Greenpeace ad warning of the dangers of genetically modified foods (cf. [48]) depicts an ear of corn as a grenade (i.e., with a grenade pin attached to the end of the corn cob). This perceptual blend, which is meant to trigger a conceptual association between GMOs and deadly weapons, is rooted in a more basic visual analogy between an ear of corn and the body of a grenade.

2.4.3 *Visual Analogy* is *"Seeing-As"*

Models such as Copycat and Metacat take seriously the idea that analogy-making is a perceptual process. However, the sort of perception being referred to—*high-level perception*—is amodal, or distinct from lower-level perceptual modalities. In this section, we have emphasized that analogy-making need not be separated from perceptual modalities; we have focused on vision, but this is not the only relevant modality, as examples involving hearing and touch could also be cited. Indeed, many cases of analogy-making and blending—from Koestler's "optical puns" to Dalí's *Mae West*—only make sense when we focus on the visual and spatial mappings that are involved. Interestingly, we find parallels not only in terms of the slippages and mappings that take place between "conceptual" and "perceptual" analogy-making, but also in the language used to talk about these examples (i.e., visual "puns," "rhymes," "metaphors," etc.), which suggests a deep connection between the two. While AI has typically emphasized the purely conceptual aspects of analogy, we feel that closer attention to visual (and other modality-specific) aspects of analogy can provide a useful lens for looking at questions of aesthetics.

2.5 Action Slips

As we saw with the MixStix example, slippage is also relevant in regard to actions and behaviors. These slippages can be intentional and playful—as when a child "plays the drums" using ordinary kitchen objects—or unintentional, as with action slips such as the following ([45], p. 95–96):

> Instead of opening a tin of Kit-E-Cat, I opened and offered my cat a tin of rice pudding.
>
> When seasoning the meat, I sprinkled it with sugar instead of salt.
>
> I raised the egg in the egg cup to my lips instead of the orange juice.

Much like Freud [19] saw slips of the tongue as portals into the "psychopathology of everyday life," more recent researchers ([40], [45]) have treated action slips as windows into the mechanisms underlying skilled behaviors, ranging from the mundane (making tea) to the highly specialized (flying an airplane). This line of research has clear practical applications, which range from improving product design to preventing disastrous errors such as airplane crashes and nuclear-power meltdowns. However, there are also interesting theoretical consequences: just as slips of the tongue reveal close ties to the processes involved in jokes, poetry, and other creative uses of language, action slips can be also striking in their seeming ingenuity.

For example, one subject in [40] reported accidentally throwing a dirty T-shirt into the toilet, rather than the intended laundry hamper, which was in another room. Another subject reported stopping his car and proceeding to unbuckle his *wristwatch* rather than his seatbelt. The perceptual and action-based slippages involved in these errors have much in common with the kinds of *conceptual* slippages that play a central role in the kind of creative analogy-making modeled in Copycat and Metacat. The *seatbelt–watchband* slip, for example, is actually supported by a clear and coherent set of mappings: from *waist* to *wrist*, from *seatbelt buckle* to *watch buckle*, and from *seatbelt webbing* to *watchband*. While the action was unintended, there is an aptness to this (unintentional) analogy that renders it worthy of appreciation, much as we might appreciate a clever joke.

2.5.1 Affordances and Slippage

The concept of *affordances*, originally put forth by psychologist J. J. Gibson [21], offers a useful framework for discussing the link between perceptual slippages and action slips. As Gibson explained, "The affordances of the environment are what it offers the animal, what it provides or furnishes, either for good or ill" (p. 127). This concept was later appropriated by HCI practitioners such as [40], who emphasized the role of *perceived* affordances in the design of artifacts. For example, when we encounter a well designed doorknob or handle, it is clear where to push or pull; no trial-and-error is necessary. In other words, the door handle's perceived affordances should reflect its *actual* affordances: It should not scream "pull me" when it needs to be pushed instead. As Norman summarized, an object's affordances "provide strong clues to the operations of things. Plates are for pushing. Knobs are for turning. Slots are for inserting things into..." ([40], p. 9).

Despite these clues, affordances can lead us astray. Action slips often result from misperceived (or carelessly perceived) affordances. On the one hand, a given object or material can afford unintended actions. Think back to the person who threw the T-shirt into the toilet: the action was *afforded* by the toilet, but certainly not *intended* by the person involved. A comical illustration of a similar slip occurs in *The Naked Gun 33 1/3: The Final Insult* when Leslie Nielsen's character—realizing he doesn't have time to make it to a toilet—regurgitates into the bell of a tuba during an upscale social gathering (much to the horror of everyone involved). On the other hand, two different objects can afford the same action. A comical example of this occurs in Mel Brooks' film *High Anxiety*, in which Brooks' character is harmlessly "stabbed" in the shower by a disgruntled bellhop, who uses a rolled-up newspaper as a "weapon." This scene is a parody of the famous shower scene in Alfred Hitchcock's *Psycho* (1960), in which the main character is fatally stabbed in a motel shower. Semantically, *knife* and *newspaper* are distant concepts; it is only because a rolled-up newspaper *affords* the action of stabbing (albeit not very effective stabbing) that this aspect of the scene makes sense.[8]

2.5.2 From Slips and Slippage to Aesthetics and Affect

The aforementioned examples illustrate the fine line that often separates inadvertent slips from creative slippages. Comedy writers play on our tendencies to commit as well as to notice (and laugh at) these types of errors. But these writers are themselves doing something creative when they concoct these scenes for our viewing pleasure. Setting these fictional and real-life examples next to one another, we can see that the difference between unintended slips and purposive, creative slippages is often simply a matter of context (e.g., fictional vs. real life) or point of view.

While we can appreciate action slips and slippages in visual art, on the movie screen, and even in studies of human error-making, they are not as desirable when we ourselves are the ones doing the slipping. Anyone who has ever turned on the wrong stove burner, only to notice an unpleasant burning smell several minutes later, can relate to the frustration that such slips can cause. The burgeoning fields of human-centered design, ergonomics, and HCI have stressed the role of good design in preventing (or at least reducing the frequency of) such frustrating slips, whether they occur in the kitchen, the workplace, or at the computer. Two salient points to emerge out of research in such fields are the importance of coherent mappings[9]—for example, between the dials and burners on a stove top—as well as the importance of ensuring that an object's perceived affordances dovetail with its actual affordances [40] (think back to the door-handle example). While these functional aspects of design were initially thought to be at odds with aesthetics, more recent research has shown a shift in viewpoints. It is now understood that good design can yield artifacts that are both user-friendly and aesthetically satisfying [41]. Reconciling the functional and aesthetic aspects of design can go a long way in facilitating positive experiences with artifacts (computational or otherwise).

To relate these ideas back to our earlier discussion of conceptual slippage and analogy-making, we would like to stress the need for convergence between affective and embodied AI, on the one hand, and analogy research, on the other. While models such as Copycat and Metacat serve as useful microcosmos in which to study the mechanisms of analogy-making, they are unable to address perception (in the lower-level, modality-specific sense of the term) or action. Furthermore, it is debatable to what extent these programs genuinely exhibit "aesthetic judgment" (at this point,

[8] The analogy between this scene from *High Anxiety* and the original scene from *Psycho* actually involves an elaborate blend of conceptual, visual, and action-based slippages, which are nonetheless effortlessly grasped by viewers who are familiar with the original scene being parodied.

[9] Recall that mappings are themselves a key aspect of analogy-making.

the scare quotes still seem necessary). To the extent that aesthetic sensibility and affect are truly embodied and intertwined with our actions in the world, it is unclear how far a disembodied, microdomain-centered approach can take us.

2.6 Conclusion

Analogy, aesthetics, and affect have all emerged as significant topics in AI and computing. Despite their close ties with one another, however, little has been done to explore this relationship. The ideas presented here are offered as an early step in such an exploration. The notion of slippage introduced here—not just in cognition, but in perception and action as well—can be seen as a unifying concept in the relationship between these three "A" words.

In line with Wittgenstein and some of his more recent commentators, we believe that "aesthetics, as a field of conceptual inquiry, should start not from a presumption that the central task is to analyze the determinant properties that are named by aesthetic predicates, but rather with a full-blooded consideration of the *activities* of aesthetic life" ([22]; emphasis in original). This view stands in sharp contrast with dominant alternatives: Unlike reductive accounts, it does not look for the source of aesthetic experience merely in the internal properties of the aesthetic object; and unlike cognitivist perspectives, it does not isolate the source in the mental processes of the experiencing subject. Rather, it attempts to locate the source in the *relationship* between subject and object as it unfolds in a given context. We find this line of thinking to be promising not only for conceptual exploration but also for the practical design of aesthetically and affectively satisfying computational systems.

By the same token, our attempt to build the discussion on models such as Copycat should be understood as being purely illustrative in character. Far from representing a genuine sense of aesthetic judgment, these programs embody a rather preliminary and stripped-down form of the kinds of connections and relationships that Wittgenstein highlighted (and that we aim to better understand). The value of these models, at least for our purposes, is their implementation in semantically sparse microdomains, which can focus our attention on the mappings and connections involved in analogy-making and aesthetic evaluation. The depth and extent of these connections in human aesthetic experience are, needless to say, much greater. Whether or not this semantic thinness and contextual sparsity would preclude AI systems from ever accomplishing that level of aesthetic engagement with their environment is a question that will be decided empirically in the future.

Acknowledgements. Our sincere thanks go to Jan Romportl and Pavel Ircing for their insightful comments, hard work, and patience over the course of this endeavor. We also thank Radek Schuster for his role in organizing the conference that led to the present volume.

References

1. Aiken, H.D.: Some notes concerning the aesthetic and the cognitive. J. Aesthet. Art Crit. 13(3), 378–394 (1955)
2. Arnheim, R.: Visual Thinking. Univ. Calif. Press, Berkeley (1969)
3. Arnheim, R.: To the Rescue of Art: Twenty-Six Essays. Univ. Calif. Press, Berkeley (1992)
4. Beardsley, M.: Aesthetics. Harcourt, New York (1958)
5. Birkhoff, G.: Aesthetic Measure. Harv. Univ. Press, Cambridge (1933)
6. Carroll, N.: Visual metaphors. In: Hintikka, J. (ed.) Aspects of Metaphor., pp. 189–218. Kluwer, Dordr. (1994)
7. Chalmers, D.J., French, R.M., Hofstadter, D.R.: High-level perception, representation and analogy: A critique of artificial intelligence methodology. J. Exp. Theor. Artif. Intell. 4(3), 185–211 (1992)
8. Danto, A.: The Transfiguration of the Commonplace. Harv. Univ. Press, Cambridge (1981)
9. Dewey, J.: Art as Experience. Perigree, New York (1934)
10. Dutton, D.: Artistic crimes: the problem of forgery in the arts. Brit. J. Aesthet. 19, 302–314 (1979)
11. Ekbia, H.R.: Artificial Dreams: The Quest for Non-Biological Intelligence. MIT Press, Cambridge (2008)
12. Evans, T.G.: A program for the solution of geometric-analogy intelligence test questions. In: Minsky, M. (ed.) Semantic Information Processing, pp. 271–353. MIT Press, Cambridge (1968)
13. Eysenck, H.J.: The experimental study of the 'good Gestalt'—a new approach. Psych. Rev. 49(4), 344–364 (1942)
14. Falkenhainer, B., Forbus, K.D., Gentner, D.: The Structure-Mapping Engine: algorithm and examples. Artif. Intell. 41, 1–63 (1989)
15. Fauconnier, G.: Conceptual blending and analogy. In: Gentner, D., Holyoak, K.J., Kokinov, B.N. (eds.) The Analogical Mind: Perspectives from Cognitive Science, pp. 255–286. MIT Press, Cambridge (2001)
16. Fauconnier, G., Turner, M.: The Way We Think. Basic Books, New York (2002)
17. Fishwick, P. (ed.): Aesthetic Computing. MIT Press, Cambridge (2006)
18. Forceville, C.: Pictorial metaphor in advertising. Metaphor Symb. Act. 9(1), 1–29 (1994)
19. Freud, S.: The Psychopathology of Everyday Life. Norton, New York (1960)
20. Gentner, D.: Structure-mapping: A theoretical framework for analogy. Cogn. Sci. 7(2), 155–170 (1983)
21. Gibson, J.: The Theory of Affordances. In: Shaw, R., Bransford, J. (eds.) Perceiving, Acting, and Knowing, pp. 67–82. Erlbaum, Hillsdale (1977)
22. Hagberg, G.: Wittgenstein's aesthetics. In: Zalta, E.N. (ed.) The Stanford Encyclopedia of Philosophy (Fall 2008 edition), `http://plato.stanford.edu/archives/fall2008/entries/wittgenstein-aesthetics/`
23. Hoenig, F.: Defining computational aesthetics. In: Neumann, L., Sbert, M., Gooch, B., Purgathofer, W. (eds.) Computational Aesthetics 2005: Eurographics Workshop on Computational Aesthetics in Graphics, Visualization, and Imaging, pp. 13–18. Eurographics Association, Aire-la-Ville (2005)

24. Hofstadter, D.R.: Fluid Concepts and Creative Analogies: Computer Models of the Fundamental Mechanisms of Thought. Basic Books, New York (1995)
25. Hofstadter, D.R.: Analogy as the core of cognition. In: Gentner, D., Holyoak, K.J., Kokinov, B.N. (eds.) The Analogical Mind: Perspectives from Cognitive Science, pp. 499–538. MIT Press, Cambridge (2001)
26. Holyoak, K.J., Thagard, P.: Analogical mapping by constraint satisfaction. Cogn. Sci. 13, 295–355 (1989)
27. Hume, D.: Essays Moral, Political, and Literary, vol. 1. Longmans, Green, and Co., Lond. (1875)
28. Irvin, S.: The pervasiveness of the aesthetic in ordinary experience. Brit. J. Aesthet. 48(1), 29–44 (2008)
29. Johannessen, K.J.: Wittgenstein and the aesthetic domain. In: Lewis, P. (ed.) Wittgenstein, Aesthetics, and Philosophy, pp. 11–36. Ashgate, Burlington (2004)
30. Kieran, M.: Revealing Art. Routledge, Lond. (2005)
31. Koestler, A.: The Act of Creation. Hutchinson, Lond. (1964)
32. Lavie, T., Tractinsky, N.: Assessing dimensions of perceived visual aesthetics of web sites. Int. J. Hum.-Comput. Stud. 60(3), 269–298 (2004)
33. Leddy, T.: The nature of everyday aesthetics. In: Light, A., Smith, J.M. (eds.) The Aesthetics of Everyday Life, pp. 3–22. Columbia Univ. Press, New York (2005)
34. Light, A., Smith, J.M. (eds.): The Aesthetics of Everyday Life. Columbia Univ. Press, New York (2005)
35. Marshall, D.: Arguing by analogy: Hume's standard of taste. Eighteenth-Century Stud. 28(3), 323–343 (1995)
36. Marshall, J.: Metacat: A Self-Watching Cognitive Architecture for Analogy-Making and High-Level Perception. Dr. Diss., Indiana Univ., Bloomington (1999)
37. Mitchell, M.: Analogy-Making as Perception. MIT Press, Cambridge (1993)
38. Moles, A.: Information Theory and Esthetic Perception. Univ. Ill. Press, Urbana (1966)
39. Norman, D.: Categorization of action slips. Psych. Rev. 88(1), 1–15 (1981)
40. Norman, D.: The Psychology of Everyday Things. Basic Books, New York (1988)
41. Norman, D.: Emotional Design. Basic Books, New York (2005)
42. Ortiz, M.: Visual rhetoric: primary metaphors and symmetric object alignment. Metaphor. Symb. 25, 162–180 (2010)
43. Petkov, G., Vankov, I., Kokinov, B.: Unifying deduction, induction, and analogy by the AMBR model. In: Proc. 33rd Annu. Conf. Cogn. Sci. Soc. Erlbaum, Hillsdale (2011)
44. Picard, R.: Affective Computing. MIT Press, Cambridge (1997)
45. Reason, J., Mycielska, K.: Absent-Minded? The Psychology of Mental Lapses and Everyday Errors. Prentice-Hall, Englewood Cliffs (1982)
46. Rehling, J.: Letter Spirit (Part Two): Modeling Creativity in a Visual Domain. Dr. Diss., Indiana Univ., Bloomington (2001)
47. Rigau, J., Feixas, M., Sbert, M.: Conceptualizing Birkhoffs aesthetic measure using Shannon entropy and Kolmogorov complexity. In: Cunningham, D.W., Meyer, G., Neumann, L. (eds.) Comput. Aesthet. Graph. Vis. Imag., pp. 105–112. Eurographics Association, Aire-la-Ville (2007)
48. Schilperoord, J., Maes, A., Ferdinandusse, H.: Perceptual and conceptual visual rhetoric: the case of symmetric object alignment. Metaphor. Symb. 24, 155–173 (2009)
49. Shelley, J.: Hume's double standard of taste. J. Aesthet. Art Crit. 52(4), 437–445 (1994)
50. Sibley, F.: Aesthetic concepts. Philosophical Review 68, 421–450 (1959)

51. Teng, N.Y., Sun, S.: Grouping, simile, and oxymoron in pictures: a design-based cognitive approach. Metaphor. Symb. 17(4), 295–316
52. Wilson, E.A.: Affect and Artificial Intelligence. Univ. Wash. Press, Seattle (2010)
53. Wittgenstein, L.: Philosophical Invesitgations. Blackwell, Oxford (1953)
54. Wittgenstein, L.: Lectures and Conversations on Aesthetics, Psychology, and Religious Belief. Blackwell, Oxford (1966)
55. Zhang, P.: Theorizing the relationship between affect and aesthetics in the ICT design and use context. In: Proc. Int. Conf. on Inf. Resour. Manag., Dubai, United Arab Emirates (2009)

Chapter 3
Rationality {in|for|through} AI

Tarek R. Besold

Abstract. Based on an assessment of the history and status quo of the concept of rationality within AI, I propose to establish research on (artificial) rationality as a research program in its own right, aiming at developing appropriate notions and theories of rationality suitable for the special needs and purposes of AI. I identify already existing initial attempts at and possible foundations of such an endeavor, give an account of motivations, expected consequences and rewards, and outline how such a program could be linked to efforts in other disciplines.

3.1 Introduction

Human behavior sometimes seems erratic and irrational. Nonetheless, it is widely undoubted that man can act rational and actually appears to act rational most of the time, making him an *animal rationale*. In explaining behavior, often terms like beliefs and desires are used: If an agent's behavior makes the most sense to us, then we interpret it as a reasonable way to achieve the agent's goals given his beliefs. This can be taken as indication that some concept of rationality does play a crucial role when describing and explaining human behavior in a large variety of situations.

Now, being faithful to the original goal of AI, namely the (re)creation of an artificially intelligent agent being at least comparable to humans in terms of intelligence and related notions, the step to considering rationality an important concept also within the context of research in artificial intelligence is a small one. If a machine should be considered as at least human-like in its capabilities and behavior, a human agent would most likely be taken as comparandum, and the alleged (ir)rationality of the machine's actions and behavior would be judged in comparison to the corresponding human examples.

Tarek R. Besold
Institute of Cognitive Science, University of Osnabrück, 49069 Osnabrück, Germany
e-mail: tbesold@uni-osnabrueck.de

J. Kelemen et al. (Eds.): Beyond Artificial Intelligence, TIEI 4, pp. 49–62.
DOI: 10.1007/978-3-642-34422-0_3 © Springer-Verlag Berlin Heidelberg 2013

Therefore, in the following, I will investigate into and reflect on the role rationality has played in the history of AI this far, will provide an assessment of the actual status quo, and will also give some proposals and recommendations for a possible future of rationality and corresponding research in relation to AI.

Sect. 3.2 offers a general introduction to previous work and theories of rationality in general, naming the four classical paradigms for modeling rationality and rational behavior, before Sect. 3.3 elaborates on rationality within the context of AI in more detail, amongst others providing motivations for work on rationality in AI, as well as an assessment of the actual situation and status quo. Sect. 3.4 then sketches the foundations of a proper research program on theories and models of rationality within AI, providing arguments in favor of such an endeavor, before Sect. 3.5 presents a vision of how foundations for a model of rationality suitable for the aims and purposes of AI might look like. Sect. 3.6 summarizes some of the main aspects of the proposed stance in a short conclusion.

3.2 Rationality

Research in rationality and rational behavior has a rich history, both within more abstract disciplines like philosophy (possibly starting with Aristotle's ascription of a rational principle to the human being in his Nicomachean Ethics [6]) or economics (for instance think of the model of the *homo oeconomicus*, conceptually at least dating back to a 1836 essay by John Stuart Mill [25]), as well as in more individual-centered fields as psychology and cognitive science (several examples are given in the following).

Over the years, many quite distinct frameworks for modeling rationality (and establishing a normative theory) have been proposed. Breaking these distinct approaches down to their underlying theoretical foundations, four main types of models can be identified, together with corresponding normative interpretations for what has to be judged rational according to the respective approach:[1]

- Logic-based models (cf. e.g. [13]): A belief is rational, if there is a logically valid reasoning process to reach this belief relative to available/given background knowledge.
- Probability-based models (cf. e.g. [20]): A belief is rational, if the expectation value of this belief is maximized relative to given probability distributions of background beliefs.
- Game-theoretically based models (cf. e.g. [26]): A belief is rational, if the expected payoff of maintaining the belief is maximized relative to other possible beliefs.
- Heuristic-based models (cf. e.g. [17]).

[1] Partially with exception of the heuristic-based models, as due to their fundamentally different approach to conceptualizing and understanding rationality classical ideas of normativity within a model of rationality become obsolete and in many cases are not anymore part of the respective frameworks.

Unfortunately, it shows that the definitions of rationality arising from within the distinct approaches are in many cases almost orthogonal to each other (as are the frameworks), making them in the best case incommensurable, if not inconsistent or even partly contradictory in their modeling assumptions. Also, in many cases the predictive power of these classical theories of rationality turns out to be rather limited (at least when applied to real-world examples instead of artificially simplified and constructed scenarios), as their emphasis clearly lies on the normative or postdictive-explanatory aspects of the respective models – which should be a crucial observation when thinking about rationality and its role and applications in AI.

Even more, although each of the listed accounts has gained merit in modeling certain aspects of human rationality, the generality of each such class of frameworks has at the same time been challenged by psychological experiments or theoretical objections:

- On the one hand, studies by Wason and Shapiro question the human ability of reasoning in accordance with the principles of classical logic [38], and also Byrne's findings on human reasoning with conditionals [8] indicate severe deviations from this classical paradigm.

  Similarly, when considering probability-based models, Tversky and Kahneman's Linda problem [36] illustrates a striking violation of the rules of probability theory.
- On the other hand, game-based frameworks are questionable due to the lack of a unique concept of optimality in game-theory. Provided e.g. with the plethora of different equilibrium concepts which have been derived from the original Nash equilibrium (cf. e.g. [21]), which one shall be taken as "“the most rational one” given a certain situation?
- Finally, heuristic approaches to judgment and reasoning follow a different approach, and are often conceptualized as approximations to a rational ideal, rather than an instantiation of the ideal itself. In some cases and application scenarios, heuristics actually do work well and can yield surprising results in practice, but still mostly lack formal transparency and explanatory power. Moreover, their status as almost magical solutions to problems relating with complexity and perceived intractability in human decision-making has recently been challenged in a quite fundamental way ([31]). But also from a more general methodological or philosophical point of view, fundamental criticism can be stated: Due to the open nature of the collection of heuristics propagated in most current accounts (i.e. whenever a phenomenon cannot be covered or described by an existing heuristics, a new one specifically fit to the task is introduced), the possibility of falsification and refutation of modeling assumptions and theories is not guaranteed, and a (reasonable) completion of the model can neither be checked for, nor feasibly assumed at any point.

3.3 A Survey: Rationality in AI

This section wants to shed some light on the concept of rationality (broadly construed), the role and importance of rationality for AI as a research endeavor, and its current standing and status within the field of AI. For the sake of the argument, I consider a very broad notion of what "rationality" is taken to be, as a limiting parameter for an action/behavior to be considered as rational only demanding for compliance with one or another variant of the quite abstract "principle of rationality" (i.e., requiring solutions to a task to constitute the least effortful, direct course of action to attain as much of a goal as it is possible given current environmental and other constraints).

In the following, initially a short perspective on the relation between rationality and intelligence in general will be given before presenting some arguments why AI actually should care about notions and theories of rationality. Concludingly, I will provide a concise overview of earlier efforts and results concerning rationality in the context of artificial intelligence.

3.3.1 Rationality & (Natural) Intelligence

In many, often folk-psychologically motivated accounts of human cognition, the notion of intelligence is mostly seen as closely intertwined with a concept of rationality. Whilst in many cases a certain form of intelligence is taken as a precondition for rationality and rational behavior, rational behavior and rationality also may be considered one possible indicator for an agent's intelligence (i.e. making some form of intelligence a necessary condition for rationality to arise). Often, both phenomena are even understood as following some direct proportionality: The more intelligent a subject is, the more rational its behavior is expected to be (and vice versa).

But here, folk psychology clearly deviates from standard theoretical accounts and models for human rationality: Whilst the latter normally do not take individual factors of the subject (as, e.g., its level of intelligence, or overall capabilities) into account, the above stated conception – by making the level of rationality also dependent on the level of intelligence – clearly does.

An account which seems to be close in spirit to the assumption underlying this non-standard notion can also be found in scholarly study of rationality, namely in the field of decision theory. In [18], Gilboa (also see an earlier definition in [19]) defines rationality as a concept that crucially depends on the subject that is executing the reasoning and (allegedly) rational behavior: "*(...) a mode of behavior is rational for a given decision maker if, when confronted with the analysis of her behavior, the decision maker does not wish to change it.*". The consequences of this seemingly simple definition are tremendous: Rationality becomes subject-centered, as now what is considered as rational might vary with the population in question. If the decision maker does not understand the analysis, or why her behavior is not judged as rational, she cannot be judged irrational for not complying with the alleged norm of rationality. If limited cognitive capacities do not allow the reasoner to understand the rules he should follow in his reasoning, but would always make him take the same decision again, he has to be called rational – although his behavior in that very moment might strike an observer as utterly irrational.

Skeptics now might be tempted to put forward the claim that, once one accepts such a notion of rationality, from that moment onwards there is no more difference between intelligence and rationality, but both notions collapse into one. To me, this objection seems mistaken: Although rational behavior might be witnessed as a (maybe even measurable) manifestation of intelligence, the effects of intelligence still do not have to be limited to rational behavior. It is commonplace that there are many different aspects to the overall notion of intelligence, an idea (amongst many others) going back to work by Howard Gardner. In [16], he lists at least eight abilities that might be considered facets of intelligence: spatial, linguistic, logical-mathematical, bodily-kinesthetic, musical, interpersonal, intrapersonal, and naturalistic. Clearly, some of the tasks and challenges addressing one or several of these abilities can also be subsumed by what a notion of rationality, built upon the aforementioned principle of rationality, can cover. Still, it should also become obvious that this coverage will not be complete, making intelligence the more general, by far more diverse and holistic conception.

3.3.2 Why Should AI Care About Rationality?

Artificial intelligence offers a wealth of concepts and notions called rational or claimed to reflect rationality. Treating rationality within a framework of artificial intelligence actually may have good reasons:

On the one hand, subscribing to a stance close to the human-like intelligence or "strong AI" research programs, modeling (human-style) rationality clearly has to be considered one of the milestones for reaching the overall goal of a "truly" intelligent AI system. Consider the example of what now is known as the "Artificial General Intelligence" (AGI) movement (cf. e.g. [37]): As explained there, an AI research project, in order to qualify for being an "AGI project", amongst others has to "*(...) be based on a theory about 'intelligence' as a whole (which may encompass intelligence as displayed by the human brain/mind, or may specifically refer to a class of non-human-like systems intended to display intelligence with a generality of scope at least roughly equalling that of the human brain/mind).*" Still humans and their performance and capabilities (although not without any alternative) stay the main standard of comparison for the targeted type of system – which in turn of course also assigns a crucial role to modeling and implementing human-style rational behavior. Similarly, a test for an artificial system's behavior in rationality tasks can serve as a crucial sub-task of a modernized decomposition [1] of Turing's famous test for machine intelligence [35].

On the other hand, staying closer to the ideas of specialized AI accounts, having a feasible concept of rationality within a system would allow agents to interact with humans in a more natural way, as well as to predict humans' expectations and interactive behavior. Applications for these capabilities would be manifold, ranging from decision support systems, through trading agents, to the use of intelligent agents in flight control scenarios. Also, from a metaperspective, AI accounts of rationality could provide inspiration, modeling tools and testbeds for theories of rationality arising within related disciplines (cf. also Sect. 3.4).

3.3.3 *Earlier Work & Status Quo*

Several prominent researchers within the field of artificial intelligence explicitly addressed the relation between AI and rationality, as well as the role rationality can play within artificial intelligence. For instance, in [32], Russell elaborated on the relation between rationality and intelligence in an AI context. Starting out from the very premise of AI itself, the assumption that the understanding and (re)creation of intelligence is possible, Russell shows the importance of having a notion of intelligence that is precise enough to serve as a basis for both, the cumulative development of general results, as well as robust systems, before he considers rational agency as a candidate for fulfilling this role (also outlining the history and development of the corresponding concepts of rationality). And also Doyle [12] – claiming that a theory of rationality might at some point equal mathematical logic in its importance for mechanizing reasoning – provides a concise survey of work at the intersection between the economic theory of rationality and AI, offering insights into how basic notions of probability, utility and rational choice (together with some pragmatic considerations) can influence the design and analysis of reasoning and representation systems.

Moreover, it should be noted that certain conceptions of rationality have played an important role at different turning points within the development of artificial intelligence as a scientific field, sometimes emerging from new paradigms within AI, sometimes directly contributing to the creation of new stances and perspectives. Following the lines of a more detailed treatment in [9] (where rationality is regarded in a very general way as "*reason-governed behavior*"), for example the following research programs and key questions with direct impact from and/or on research in rationality can be named:

- Robotics: Can reasoned action be explained without making appeal to inner, form-based vehicles of meaning (i.e. are internal representations perhaps superfluous)? And in consequence, can there be something like representation-free rationality – can phenomena like what we consider "deliberative reasoning" or "abstract thought" be explained by a complex of reflex-like mechanisms alone?
- Global reasoning: How can "non-classical" forms of reasoning (e.g. abductive or analogical reasoning) which are clearly present in humans as real-world rational agents be accounted for in AI systems (where still deductive reasoning as the dominant paradigm)?
- Heuristics: Given the already aforementioned conceptions of fast and frugal heuristics [17] as real-world mechanisms of decision-making and reasoning, how must the principal ideas and understanding underlying most approaches to mechanizing reasoning (and thus also to mechanical rationality) be altered and adapted? Which techniques and AI paradigms are best fit to implement such heuristics-based mechanisms?

Concerning the status quo at the present moment, the different families of models for rationality listed in Sec. 3.2 (partly with exception of the heuristics-based approach) can actually be found within existing AI systems and theories. Still, when having a closer look at the latter, it shows that the underlying notions of rationality have stayed close to their fields of origin, having been adapted only to a minimal degree (if at all). Also, possible deficits and shortcomings have been brought along without questioning or fixing.

In consequence, AI systems e.g. mostly fall short in tasks such as predicting or exhibiting behavior resembling human-like rationality, a crucial need in all domains concerned with close interaction/cooperation between a human user and an AI system. Here, examples are numerous, ranging from rational agents communication for cooperative dialogues [34] to adaptive and cooperative wheelchairs [15].

3.4 The Goal: Rationality for AI

Recently, some researchers in cognitive science and decision theory are questioning the completeness and suitability of the classical approaches to rationality. For example, having a position not that different from Gilboa's view (sketched in Sect. 3.3.1), also Kokinov challenges traditional views on rationality in [23]: Initially observing that rationality fails as both, descriptive theory of human-decision making and normative theory for good decision-making, Kokinov concludes that the concept of rationality as a theory in its own right ought to be replaced by a multilevel theory based on mechanisms and processes involved in decision-making. He demands the classical concept of utility making to be rendered as an emergent property (as should the concept of rationality itself), emerging in most (but not all) cases.

One possible consequence of this stance is the proposition to use humans as gold-standard for actually existing rational agents, and consequently base models of rationality and rational behavior on cognitive capacities (as, for instance, analogy-making, cf. e.g. [2]). Clearly, this brings along a fundamental shift in the type of theory, aiming not for theories of normative nature, but trying to build a positive theory of human rationality (as e.g. also mentioned in a side remark in [5]): Until now, almost every theory of rationality put its emphasis on providing a framework for postdictively deciding whether an already taken action or decision ought to be considered rational or not. Contrastingly, a positive theory would focus on the predictive part of its model of rationality, aiming at feasibly predicting a rational agent's behavior and decisions when being provided with a precise-enough description of all information relevant for the respective situation (a stance which from my point of view also seems way closer to application scenarios and settings within AI).

First theoretical and empirical studies on using computational cognitive systems for these modeling tasks seem to provide support for the practical possibility and valuable applicability of approaches aiming at also taking cognitive aspects of humans into account when creating models of (human-style) rationality: In work on the JUDGEMAP system [27] for judgment and choice, particularities and characteristics which normally are taken as typical for humans could be reproduced in the system's behavior, and considerations concerning applications of the computational analogy engine HDTP to rationality tasks (cf. e.g. [3]) offer new solution strategies for long-standing challenges for the standard models of rationality.

Following these examples, additionally being inspired from within AI e.g. by [33], I argue for establishing work on rationality as a proper, coordinated research program within artificial intelligence, not only limiting its focus to taking theories of rationality from respective neighboring disciplines and adapting them so that they can be applied within an AI framework, but to actively pursue research on (artificial) rationality on its own.

The advantages of such an approach seem significant: From an AI-internal perspective, it allows for approaching rationality within artificial intelligence from a more holistic point of view, possibly amalgamating existing accounts into theories and conceptions more suitable for the use in artificially intelligent systems. Here, early examples can, e.g., already rudimentarily be found in the development of probabilistic dynamic epistemic logics [24] (an attempt at integrating probabilistic and logical views of rationality), or in the conception of "algorithmic rationality" [22], which offers a framework for including computational costs in otherwise game-theoretical notions of rationality. The latter also may serve as an example for a second advantage of a properly AI-oriented notion of rationality: Most classical models and theories of rationality do not take into account computationally crucial factors as e.g. the complexity of a reasoning procedure. This might still be justifiable and legitimate in a field like economics or maybe even psychology, but shows to be a major problem when talking about rationality in an AI context. As AI this far necessarily and always is dealing with some implementation of a Turing machine, complexity issues do play a crucial role when implementing and emulating any kind of intelligence or cognitive capacity at any level, and the tractability of the used methods and algorithms becomes a key issue (for a related perspective from the field of Cognitive Science, cf. also [30]). When establishing work on rationality as a proper program within AI, one intuitive starting point would therefore be to revisit the traditional accounts and models of rationality developed over the decades, and perform a rigorous complexity and tractability analysis on them. Doing this in a coordinated, collaborative way within an overall research program can reasonably be expected to be beneficial in several ways, amongst others allowing for the re-use of results and insights between different sub-endeavors, the immediate uptake of new ideas and conceptions into new models, frameworks and possibly even paradigms by neighboring projects and groups within the overall rationality AI program, but also creating the possibility to directly test and evaluate newly constructed models and theories against the then existing benchmarking infrastructure (on both, a conceptual and a computational level).

Finally, taking all these aspects into account, the proposed approach would thus also allow for providing feedback to neighboring disciplines, directly guiding further research in rationality concepts for AI, and could thereby play an integrative role similar to the artificial general intelligence program within the development of "strong AI" (cf. e.g. [37]).

From a more high-level point of view, it gives a basis for closer cooperation with neighboring fields, allowing for more and deeper (also not only unidirectional, but bidirectional) interaction. Such joint efforts might span a very wide range of scenarios:

- Concrete project-based joint ventures for example in human-computer interaction: In [7], Butterworth and Blandford state that an approach to reasoning about interactive behavior can be based on the assumption that computer users are rational, and that the behavior of the interactive system as a whole results from the rational behavior of the users in combination with the programmed behavior of the system's parts. Of course, the reliability of such an approach crucially depends on the quality of the rationality model, and the latter's suitability for such a use – where specifically designed models of rationality, deviating from the classical paradigms and (up to a high degree) reliably predicting human rational behavior, can be expected to have major benefits and advantages over the classical postdictive theories and accounts.
- Rather specific applications of AI methods and systems for mitigating shortcomings of theoretical frameworks in economics: In [11], whilst still maintaining a rather orthodox view on economic rationality theory, Dixon also elaborates on possible contributions of the field of artificial intelligence to the study of rationality. He sketches two examples where AI might play a significant role, one of them being disequilibrium situations (i.e. situations in which agents in their behavior deviate from the classical notion of equilibrium and thus seemingly err in their actions), and the second one being strategic environments (i.e. situations and contexts in which agents can increase their obtained utility by behaving seemingly non-optimal). Whilst economic theory faces great difficulties in modeling both phenomena, Dixon is rather confident that AI-inspired techniques could offer new insights and results.
- Inspiration, modeling tools and testbeds for theories of rationality within other disciplines: In [14], Frantz traces the origins of Herbert Simon's ideas about limited or bounded rationality back to Simon's work in the field of AI, and provides evidence that the former only could come about on the basis of Simon's previous experiences with the latter.
- Direct new contributions to longstanding concepts of rationality, as for instance the already mentioned notions of algorithmic rationality [22] or probabilistic dynamic epistemic logic [24].
- Fundamental research on human-like rationality conducted together with researchers on the cognitive science-side of system design and modeling, epistemology and philosophy of cognition: Here, Pollock's OSCAR Project (cf. e.g. [28]) can serve as an example. The project's aims were twofold, on the one hand, a general theory of rational cognition should be constructed, on the other hand, also an artificial rational agent implementing that theory should be built, resulting in a joint project in (at least) philosophy of cognition and AI. The basic idea of this particular approach was to offer the philosophical side a possibility of testing the correctness of the corresponding theory of rational cognition by applying it to concrete examples via the agent, whilst AI should be provided with a

general characterization of what it is to make rational decisions and draw rational conclusions (i.e. a general theory of rational cognition) from the philosophical side.

The success (or lack thereof) of such a rationality program within AI could at least be measured against two dimensions, namely the overall impact within AI and the impact within the classical "rationality disciplines" (i.e. psychology, economics, cognitive science). If the outcomes of associated research endeavors would contribute to coming closer to a proper and more natural (in the sense of flexible, adaptable or behaviorally adequate) model of rational behavior within artificial agent populations, or should allow for artificial systems predicting human decision-making and judgment in a more reliable and feasible way than actual architectures do, certain success of the research program could not be neglected. And also concerning the impact and consequences for topically related disciplines, a similar argument can be made: If insights resulting from the work done in AI (as, e.g., concerning the already mentioned complexity studies of existing models of rationality) can help clarifying open questions concerning the suitability and applicability of existing frameworks and theories, or can contribute to constructing better and maybe even more cognitively adequate models, the efforts spent would have to be judged worthwhile. And, although possibly more complicated to measure, also a third dimension of evaluation might be considered, namely the bridging of gaps and the creation of new connections between AI and some of its conceptually related neighbors in the world of academic disciplines, for example re-establishing some more of the original (programatic, conceptual and even methodological) links between AI, cognitive science and (cognitive) psychology.

3.5 An Outlook: Rationality Through AI

In this penultimate section, I very briefly want to sketch a few essentials and guidelines of what I consider a concept and theory of (human-style) rationality to which AI could contribute greatly (a more detailed elaboration can for example be found in [4]). This notion fundamentally deviates from most "standard" accounts of rationality, as it aims at a positive rather than normative type of model in the first place (deriving the normative dimension from the positive one once the latter has been constructed) and does not make any a priori commitments to particular formalisms or modeling techniques.

From my point of view, rationality in a human context clearly has to be considered as a subject-centered notion, also taking into account the particular capabilities and limitations of the respective agent. This position also includes the claim that there is no use establishing models and norms which can never be implemented or fulfilled by human reasoners due to limitations of the agent or of its environment (again, cf. e.g. [30]). Also, the main aspect of theories and models of rationality should be their use and applicability as positive theories, and not as mere normative or postdictive explanatory accounts. Such a model is intended to provide a reliable prediction of human rational behavior and decision-making when being supplied with the information the human reasoner can access at the moment of reasoning (and with that information only).[2] Thirdly, there does not have to be any kind of commitment to a particular modeling paradigm or technique, but an integrative approach including and integrating different formalisms and approaches to modeling is perfectly admissible.

Provided with these guiding principles, it should become clear why AI can and should play an important role in the development of the respective theory of rationality. All three major demands, i.e. the awareness of the limits and boundaries of an agent, the request for applicability as predictive model, and the acceptance of systems which integrate different formal approaches and paradigms, are already popular (if not even commonplace) in important parts of AI as a field. Therefore, this expertise and experience could be very beneficial in the development of such a positive, subject-centered notion of rationality, whilst on the other hand AI itself could greatly profit from being provided with results and theoretical outcomes from this endeavor.

I want to conclude this section with a short statement about what rationality in AI from my point of view does not have to be or provide: Theories of rationality, although hopefully being of the type just sketched in the previous paragraph, do not have to re-construct or re-create human rationality (or even human intelligence, bringing along rationality as a by-product) on a level similar to the "algorithmic level" in Marr's Tri-Level Hypothesis of information processing [10], but can stay limited to the "computational level". Although inspiration might be taken from how humans actually perform rational decision-making and judgment, insights from respective studies should not have mandatory character, but are at most to be taken as recommendations (that can, and most likely will, be ignored in quite some cases). This conception also brings along a perspective different from the basic approach to AI that seems to underlie the thoughts presented in Chap. 4 of this book. Whilst some researchers take it as a necessity to use the proper "human model" and the proper "human implementation", thus trying to re-implement particularly human faculties and capacities when trying to build an AI (seemingly aiming for a complete match at least on the computational level and on the algorithmic level in Marr's hierarchy), I take a stance much closer to the intuitions also shining through in the Turing Test [35]. For me, the important part is to get the behavior right, and not so much the way this comes about – or, quoting a philosopher friend, phrasing it in an overly simplified but quite appealing way: "*Rationality is, what rationality does.*"

3.6 Conclusion

We have now seen a short assessment of the history and the actual status of the concept of rationality within AI, have considered the possible merits, value and

[2] My approach at this point has close connections to many of the ideas and insights underlying the notion of "ecological rationality" ([29]) discussed in economics and psychology. Also, both accounts share the general fundamental criticism concerning the classical normative approach to rationality.

consequences of work on models and theories of rationality using AI techniques and methods, and have also had a quick look at a non-standard (positive instead of normative) notion of rationality that seems more useful to the purposes of AI than most classical accounts do.

Summarizing, I once again want to put forward the claim and demand that targeted research on theoretical and applied aspects of rationality within AI should be conducted, using methods from AI and, in the long run, aiming at solving some of the fundamental problems the overall artificial intelligence program tries to answer. Whilst research on rationality has – for historical, but also for social reasons within scientific communities – by now mostly been conducted almost exclusively in the fields of economics, decision theory and psychology, I am convinced that a proper research program within AI could be fruitful not only for the purposes of AI itself (here also allowing for new approaches and takes on long-standing challenges), but also would provide valuable feedback and new perspectives for the already established "players in the field of rationality".

Acknowledgements. The author wishes to thank Maricarmen Martínez Baldares for her helpful recommendations and comments concerning some of the mentioned topics. Also, thanks go to Frank Jäkel and the members of the AI Research Group at the Institute of Cognitive Science for quite some valuable discussion and exchange of ideas. Finally, I owe a debt of gratitude to Wendy Wilutzky for coining the concluding quote of Sect. 3.5.

References

1. Besold, T.R.: Turing Revisited: A Cognitively-Inspired Decomposition. In: Müller, V.C. (ed.) Theory and Philosophy of Artificial Intelligence. SAPERE (to appear, 2012)
2. Besold, T.R., Gust, H., Krumnack, U., Abdel-Fattah, A., Schmidt, M., Kühnberger, K.: An Argument for an Analogical Perspective on Rationality & Decision-Making. In: van Eijck, J., Verbrugge, R. (eds.) Proceedings of the Workshop on Reasoning About Other Minds: Logical and Cognitive Perspectives (RAOM 2011), CEUR Workshop Proceedings, Groningen, The Netherlands, vol. 751, CEUR-WS.org (2011)
3. Besold, T.R., Gust, H., Krumnack, U., Schmidt, M., Abdel-Fattah, A., Kühnberger, K.U.: Rationality Through Analogy - Towards a Positive Theory and Implementation of Human-Style Rationality. In: Troch, I., Breitenecker, F. (eds.) Proc. of MATHMOD 12, Vienna (2012)
4. Besold, T.R., Kühnberger, K.U.: E Pluribus Multa In Unum: The Rationality Multiverse. In: Proceedings of the 34th Annual Conference of the Cognitive Science Society, Cognitive Science Society, Austin (2012)
5. Binmore, K.: Rational Decisions. Princeton University Press (2011)
6. Broadie, S., Rowe, C. (eds.): Aristotle Nicomachean Ethics: Translation, Introduction, and Commentary. Oxford University Press (2002)
7. Butterworth, R., Blandford, A.: The principle of rationality and models of highly interactive systems. In: Sasse, M.A., Johnson, C. (eds.) Human-Computer Interaction – INTERACT 1999, IOS Press (1999)
8. Byrne, R.: Suppressing valid inferences with conditionals. Cognition 31(1), 61–83 (1989)

9. Clark, A.: Artificial Intelligence and the Many Faces of Reason. In: The Blackwell Guide to Philosophy of Mind. Blackwell (2003)
10. Dawson, M.: Understanding Cognitive Science. Blackwell Publishing (1998)
11. Dixon, H.: Some Thoughts on Economic Theory and Artificial Intelligence. In: Surfing Economics: Essays for the Enquiring Economist, Palgrave (2001)
12. Doyle, J.: Rationality and its roles in reasoning. Computational Intelligence 8(2), 376–409 (1992)
13. Evans, J.: Logic and human reasoning: An assessment of the deduction paradigm. Psychological Bulletin 128, 978–996 (2002)
14. Frantz, R., Simon, H.: Artificial intelligence as a framework for understanding intuition. Journal of Economic Psychology 24, 265–277 (2003)
15. Galluppi, F., Urdiales, C., Sandoval, F., Olivetti, M.: A study on a shared control navigation system: human/robot collaboration for assisting people in mobility. Cognitive Processing 10(2), 215–218 (2009)
16. Gardner, H.: Frames of Mind: The Theory of Multiple Intelligences. Basic Books, New York (1983)
17. Gigerenzer, G., Hertwig, R., Pachur, T. (eds.): Heuristics: The Foundation of Adaptive Behavior. Oxford University Press (2011)
18. Gilboa, I.: Questions in decision theory. Annual Reviews in Economics 2, 1–19 (2010)
19. Gilboa, I., Schmeidler, D.: A Theory of Case-Based Decisions. Cambridge University Press (2001)
20. Griffiths, T., Kemp, C., Tenenbaum, J.: Bayesian Models of Cognition. In: The Cambridge Handbook of Computational Cognitive Modeling. Cambridge University Press (2008)
21. Halpern, J.Y.: Beyond Nash Equilibrium: Solution Concepts for the 21st Century. In: Proc. of the 27th Annual ACM Symposium on Principles of Distributed Computing (2008)
22. Halpern, J.Y., Pass, R.: Algorithmic rationality: Adding cost of computation to game theory. ACM SIGecom Exchanges 10(2), 9–15 (2011)
23. Kokinov, B.: Analogy in decision-making, social interaction, and emergent rationality. Behavioral and Braind Sciences 26(2), 167–169 (2003)
24. Kooi, B.P.: Probabilistic dynamic epistemic logic. Journal of Logic, Language and Information 12, 381–408 (2003)
25. Mill, J.S.: On the definition of political economy, and on the method of investigation proper to it. In: Essays on Some Unsettled Questions of Political Economy, 2nd edn. Longmans, Green, Reader & Dyer (1874)
26. Osborne, M., Rubinstein, A.: A Course in Game Theory. MIT Press (1994)
27. Petkov, G., Kokinov, B.: JUDGEMAP - integration of analogy-making, judgement, and choice. In: Proceedings of the 28th Annual Conference of the Cognitive Science Society (CogSci), pp. 1950–1955 (2006)
28. Pollock, J.: Twenty Epistemological Self-profiles: John Pollock (Epistemology, Rationality and Cognition). In: A Companion to Epistemology, pp. 178–185. John Wiley and Sons (2010)
29. Rieskamp, J., Reimer, T.: Ecological rationality. In: Encyclopedia of Social Psychology, pp. 273–275. Sage, Thousand Oaks (2007)
30. van Rooij, I.: The tractable cognition thesis. Cognitive Science 32, 939–984 (2008)
31. van Rooij, I., Wright, C., Wareham, H.T.: Intractability and the use of heuristics in psychological explanations. Synthese (2010) (published online: November 11, 2010)
32. Russell, S.: Rationality and intelligence. Artificial Intelligence 94, 57–77 (1995)

33. Russell, S.J., Wefald, E.: Do the right thing - studies in limited rationality. MIT Press (1991)
34. Sadek, M.D., Bretier, P., Panaget, F.: ARTIMIS: Natural dialogue meets rational agency. In: Proceedings of IJCAI 1997, pp. 1030–1035. Morgan Kaufmann (1997)
35. Turing, A.: Computing Machinery and Intelligence. Mind LIX (236), 433–460 (1950)
36. Tversky, A., Kahneman, D.: Extensional versus intuitive reasoning: The conjunction fallacy in probability judgement. Psychological Review 90(4), 293–315 (1983)
37. Wang, P., Goertzel, B.: Introduction: Aspects of Artificial General Intelligence. In: Goertzel, B., Wang, P. (eds.) Advances in Artificial General Intelligence - Proc. of the AGI Workshop 2006, Frontiers in Artificial Intelligence and Applications., vol. 157, pp. 1–16. IOS Press (2007)
38. Wason, P.C., Shapiro, D.: Natural and contrived experience in a reasoning problem. The Quarterly Journal of Experimental Psychology 23(1), 63–71 (1971)

Chapter 4
Usage of "Formal Rules" in Human Intelligence Investigations

Tzu-Keng Fu

Abstract. In this paper, we challenge the anti-mechanism in the philosophy of mind and artificial intelligence by explaining that human intelligence, expertise, and skills acquaintance, relying mainly on the unconscious instinct could have been taken in formal rules. I do so, showing thereby the usage of the formal rules in unconscious processes for the study of human intelligence was raised in the second half of the 20th century by psychoanalyst Matte Blanco. In this way, by exploring the bi-logic framework with the general rules of unconscious processes, some positions that anti-mechanism has already taken should be re-examined in an interdisciplinary manner.

4.1 Introduction

Traditionally, the unconscious processes investigations might have been thought of as a privilege of psychoanalysts, psychotherapists, philosophers, etc. However, to taking modern logic to support some interdisciplinary studies between unconscious and logic has been made by Matte Blanco. Just as in the field of non-classical logics, he draw on a small array of formal logics with varying the domains within the Sigmund Freud's theories of unconscious processes and directed toward a different framework to characterize the unconscious and conscious processes. We here explore the interplay between unconscious processes and logic, the characteristics they have, and ways with which logic can be combined ([6], [7], [34]).

More specifically, this paper is about the intersections between the main characteristics in unconscious processes and some modern formal methods. It is not, however, about therapy, nor does it discuss the psychoanalysis. Rather, we outline a theoretical environment to discuss the formal characteristics of unconscious processes.

Tzu-Keng Fu
Department of Computer Science, University of Bremen, Bremen, Germany
e-mail: tzukeng.fu@gmail.com

J. Kelemen et al. (Eds.): Beyond Artificial Intelligence, TIEI 4, pp. 63–74.
DOI: 10.1007/978-3-642-34422-0_4 © Springer-Verlag Berlin Heidelberg 2013

The structure of the paper is as follows: In Section 2, we present the bi-logic framework based on the Matte Blanco's theory of mind and introduce the concept of "mental depth" with respect to the processes of symmetrization. We here introduce five stratum used in the Matte Blanco's theory of mind, and discuss their varying *unconsciousness*. Section 3 then discusses a number of physical features of unconscious processes. Section 4 discusses the logical features of unconscious processes, including inconsistency, explosion, and paraconsistency that have already taken as the formal properties in modern science, in particular, we turn our attention to the area of non-classical logics. Finally, section 5 summarizes the paper and discusses future work.

4.2 From Freud to Bi-logic Framework

The study of unconscious processes can be carried out to a quite large extent that Freud's theory has used in psychoanalysis. Generally speaking, Freud's greatest contribution was probably to show the importance of unconscious processes and to formulate five main characteristics of unconscious processes: the timelessness, the replacement of external by internal reality, the condensation, the displacement, and the absence of mutual contradiction (compare e.g. [7], [16], [17], [34]).

The characteristics of unconscious processes was re-explained by Matte Blanco in the later 1970s. He captured a very abstract and flexible notion of logical system to describe the unconscious processes, moreover related the psychoanalytic studies to this logical system, namely the bi-logic framework (see e.g. [6], [7], [8], [34]).

The insight of bi-logic viewpoint shows the importance of the symmetrization in mental processes. In this way, the part-whole identity and the experiences of infinity, observed in psychoanalysis practices can be fully explained. Blanco followed Freud and shew how the main characteristics of unconscious functioning can be seen as arising out of symmetrized thought. In the following, we describe the bi-logic framework to illustrate the characteristics in the unconscious processes.

4.2.1 Two Principles in the Bi-logic Framework

4.2.1.1 The Principle of Generalization

It is the first principle in bi-logic framework. It says that any individual thing (person, object, concept) is treated as if it was a member of a class (set) which contains other members. The class is further treated as a subclass of a more general class, and so on.

It is one function of the unconscious processes to look for the similarities with the classification. It has been treated as one peculiarity of the unconscious processes. It states that people abstract a particular thing to a general level in terms of the *class membership*, i.e., leap upward to the more general level of the class hierarchies in the unconscious processes.

4.2.1.2 The Principle of Symmetry

It is the second principle in bi-logic framework. It says that some relation is selectively identical with its converse. More specifically, the unconscious processes *selectively* treat the logically asymmetric relations as if they were symmetric ([7], p. 38).

There are some examples of asymmetric relation as follows:

Example 4.1. Some A is before some B has the converse of some B is after some A, and some C is the mother of some D has the converse of some D is the child of some C. Here, we say that there is an asymmetric relation between A and B with respect to the "before-after" relationship. Similarly, there is another asymmetric relation between C and D with respect to the "mother-child" relationship.

There are some examples of symmetric relations as follows:

Example 4.2. Some A is near some B has the converse of some B is near some A, and some C is the sibling of some D has the converse of some D is the sibling of some C. Here, we say that there is a symmetric relation between A and B with respect to "near-near" relationship. Similarly, there is another symmetric relation between C and D with respect to "sibling-sibling" relationship.

The asymmetric relation gains an importance from its function of locating the self within the physical world. The symmetric relation gains an dual-importance from it function of transforming or rotating an identity between different states in the physical world.

4.2.2 The "Logic" of Unconscious Processes

Following Freud's theory, Matte Blanco examines the characteristics of unconscious, and he concludes that the unconscious processes follow something non-conventional style than that in the conscious processes.

> "Matte Blanco concluded that the mind can usefully be conceived as partly functioning by the combination of at least two distinct modes of knowing which are often polarized." ([34], p. 1)

First of all, to describe highly complex dynamic mental structures, Blanco followed Freud's analysis of unconscious in terms of the interactions between some precise concepts in mathematics, including the concepts of sets, numeration, symmetry, asymmetry, dimension and infinity, to produce the "logic" of unconscious processes. Secondly, he proposed a combination of the conventional logic with the "logic" of unconscious to describe the overall mind, namely the *bi*-logic framework (compare e.g. [6], [7], [20], [34]).

Readers have to note that we are talking about the bi-logic framework in the general level of logic. Each individual logic of the *bi*-logic will be specified, if we are downward from the general level to the concrete level. With respect to the conventional logic that has already been referred to classical logic by Matte Blanco (compare e.g. [6], [7], [34]), the "logic" of unconscious processes is an augmentation of classical logic by adding two extra principles: the principle of generalization and the principle of symmetry. In this way, the concept **mental depth** was proposed as follows:

The whole mind is continuously stratified with respect to the degree of symmetrization. One side of the spectrum of the mind is about the unemotional physical world, the other side is about the unbounded emotional psychic world. As described by Matte Blanco's theory of mind, there are five stratum of the mental depth that range from unemotional physical world (asymmetric) to the emotional psychic world (symmetric):

- the conscious and well-delimited objects stratum,
- the more or less conscious emotions stratum,
- the symmetrization of the class stratum,
- the formation of wider classes which are symmetrized stratum,
- the mathematical limited stratum

This stratified structure represents "a model of the whole mind at work" to look at the mind overall. There is a similar opinion proposed by Eric Rayner:

> "The stratified structure is Matte Blanco's major essay into creating a model of the whole mind at work rather than the functioning of aspects of simply one way of looking at the mind overall. It is different from both Freud's id, ego, super-ego model and, for instance, from Fairbairn's endopsychic structure; but it is not intended to be a replacement for either." ([34], p. 77)

4.3 The Physical Features of Unconscious Processes

According to the bi-logic framework, to study the unconscious processes by formal methods becomes potentially reliable. As we have seen, Matte Blanco's theory of mind sets up a two poles structural analysis of mind by taking the bi-logic framework. It explains the five main Freudian characteristics of unconscious processes: *the timelessness, the replacement of external by internal reality, the condensation, the displacement, the absence of mutual contradiction*, as arising out of the symmetrization. In this section, we pin down each of these characteristics.

4.3.1 The Conceptions of Space-Time Taken in Matte Blanco's Theory of Mind

4.3.1.1 The External Space

In Matte Blanco's theory of mind, the physical external space was not referred to the Newtonian absolute space but much closer to the Einstein's conception of space-time. It is a system of relations in spatial dimensions (three-dimensions) with time that is of a different sort from the spatial dimensions.

4.3.1.2 Psychological Space

In Matte Blanco's theory of mind, the psychological space was defined as the intuitive notion of the external space. More specifically, this internal space was referred to the "inner place" that experiences reside. Such a conception of psychological space based on the anthropocentric thinking.

4.3.1.3 Mathematical Space

In Matte Blanco's theory of mind, the mathematical space was conceived as an "ideal space" that usually employed in the scientific studies about the physical world. The mathematical space is an approximate description of the physical space with three basic dimensions (length, width, height). It should be treated as achieving the generality that suitable for the deductive thinking.

4.3.2 The Dimensional Transformation between the Conscious Processes and Unconscious Processes

It was observed that the unconscious processes have the capacity of multi-dimensional representations. A particular event happened in the external space can be represented to an abstraction expression (compare e.g. [6], [7], [34]). For example, it is straightforward that some events in external space can be found easily in the dreams, where people view the dream situations as the unconscious processes.[1]

The capacity of representation in conscious processes is 3 + 1 dimensional. If someone tries to have some interpretations of dreams, it implies that he is trying to represent the abstract expressions in the unconscious processes within the conscious processes. The abstract expressions in dreams will be replaced by some specific visual or pictorial expression that are able to be represented as the images of conscious processes. A mathematical notion that some relatively higher dimensional spaces can be represented as some lower dimensional spaces will help us to realize the dimensional transformation between unconscious processes and conscious processes. To take the triangle ABC as an example, an triangle is a two-dimensional surface consists of three lines $\overline{AB}$, $\overline{BC}$, and $\overline{CA}$, respectively. If we consider representing triangle ABC as a straightline in one-dimension, then it is necessary that some point will be the repetition, e.g. $\overline{ABCA}$.

Yet, the multi-dimensional reduction could be used to explain some basic facts described in the psychoanalysis. Some clinical examples can be found in [34], pp. 87-92.

[1] "Sigmund Freud saw dreams as the 'royal road to the unconscious' whose bizarre character was due to censorship and disguise of thwarted drives [...]" ([22], p. 230).

4.3.3 *The Infiniteness of Emotion in Psychotic*

Reviewing two principles that have been added to the conventional logic by Matte Blanco, we should say that the principle of generalization endorses some conception of set theory, and the principle of symmetry is about the logical system itself. These two principles, in their combination, perfectly contribute to the psychoanalysis practices.

According to Matte Blanco's theory of mind, the experience of infinity in psychoanalysis refers to a certain sense of part–whole identification. The part–whole identification, deriving from the principles of generalization and symmetry, is the underlying process of the *infiniteness* in the unconscious processes. By associating with a notion in set-theory that the part–whole equivalence between an infinite set and its subset, Matte Blanco moreover treats the "unconscious as the infinite sets" ([7]).

In psychoanalysis, the emotion was seen as rooted in the unconscious processes. The emotional states and the degree of unconsciousness were influenced with each other. Thus, the highly emotional and irrational states lie near to the unconscious states, albeit they are still with some conscious content. As described by Matte Blanco, these observations about the unconscious processes in psychoanalysis could be illustrated by the abovementioned two principles in the bi-logic framework. To put the idea of symmetrization together with infiniteness, some unconventional logic was perceived as some "reasoning" about emotion. For example, the extremely fear, anger, hate, love, etc. which are expressed by the patients will appear unlimited.

4.3.4 *The Characteristics of Space-Time in the Unconscious Processes*

The unlimited (infinite) emotion can have been expressed or rooted as a psychological consequence of symmetrization. Some physical consequences could have been formulated in the same way, including *timelessness*, *displacement*, *condensation*.

To the extend that the symmetrization implies *timelessness*, given one event A comes **after** the other B (or event B comes **before** A), by the symmetrization, it follows that event B is after event A (or event A comes after B). The terms "after" and "before" lost their canonical meanings. There is no *time* presented in serial. Similarly, following the timeless situation, the *spaceless* is simultaneously presented to make every spatial terms meaningless.

To the extend that symmetrization implies *condensation* and *displacement*, it leads to an unusual spatial geometry in the unconscious processes. These two characteristics could have been found in the dreams. For example, somebody dreamed of his different periods of life at once, both his child life at the primary school and his adult life in the university. This example accompanies with not only timeless but also the condensation. By the principle of generalization, both the child-life event and adult-life event are treated as members of the general situation, e.g. "at the school", further by the principle of symmetry, the child-life event belongs to the general situation implies that the general situation belongs to the child-life event. The adult-life event behaves in the same way. Thus, the equivalence holds between the child-life event, adult-life event, and the general situation. The condensation phenomenon occurs.

The displacement is a similar phenomenon. People may evoke some feeling felt toward another. Taking the following facts as examples:

- Your mother feeds you.
- Your professor feeds you the knowledge.

The former belongs to the more general class of *someone who feeds*, and the latter belongs to the class of *someone who feeds mentally* that belongs further to the general class of *someone who feeds*. It is the displacement that someone felt the professor feed him like the mother did.

Finally, it is straightforward that the feature: the replacement of external by internal reality could be treated as the consequence of the spacelessness.

4.4 The Logical Features of Unconscious Processes

The rational people proportion their beliefs in the conscious processes. The consistency, no doubt, indicates a fundamental criterion of specifying conscious processes and some states of rationality. Obviously, the unconscious processes are in lack of a portion of consistency and rationality. What role does, or should, inconsistency play in the unconscious processes? The question was hardly a new one, yet the development of modern non-classical logics has provided the background, if we tried to discuss the logical properties in this perspective.

4.4.1 The Absence of Classical Negation

When acknowledging the situation that there is the absence of mutual contradiction in the unconscious processes, it implies that each underlying "logic" of various stratified unconscious processes should not be the traditionally bivalent. More specifically, it implies there is no "classical logic negation" in the unconscious processes. The classical negation behaves as follows: if we assign the true value to a proposition p, then the classical negation of p will have the false value; if we assign the false value to p, then the classical negation of p will have the true value. Generally speaking, in classical logic, the conjunction of a proposition and its negation should be a contradictory proposition with the false value.

In order to understand the "negation" better, let us discuss it in a bit more detail. An abstract structural bi-logic framework consists of various "logical" modes for different mental depth, where the grades of symmetrization appear differently. Typically, the more someone appears closer to the consciousness with less symmetrization, the more significant negation mechanism he will equip with; the more someone appears closer to the unconsciousness with more symmetrization, the less significant negation mechanism he will equip with. In this way, the symmetrization runs between them to weaken the classical negation mechanism to some kinds of non-classical negations with a gradual of decrease. In other words, the logic of each mental depth is decorated with its non-classical negation.

4.4.2 The Paraconsistency

To be able to write down such bi-logic framework in a precise manner, we use the typical way taken in modern formal logic. Traditionally, *the principle of explosion (ex contradictione sequitur quodlibet)* (ECSQ) is admitted in classical logic. ECSQ means "from a contradiction, anything follows". Or we can see it as a combination of inconsistency (two contradictory propositions) and explosion (any proposition follows from the contradiction).

In Matte Blanco's theory of mind, the conventional logic refers to the classical logic. With the gradual of decrease of asymmetrization, the conscious processes gradually close to the threshold, after which would enter into the unconscious processes. The feature that the absence of mutual contradiction in unconscious processes could be interpreted as the absence of classical negation in classical logic ([34], p. 45). However, it does not mean the mind never knows "negation". Here is such a voice:

> [...] he did not invent the way the unconscious levels of the mind work. Nor is he precisely proposing a mind that *never* knows negation or contradiction anywhere – very definitely not. ([34], p. 46)

As we have seen, the processes between unconscious and conscious accompany with the *selectively localized symmetrization*. Thus, the negation accordingly appears selectively between the conscious processes and the unconscious processes. On the side of conscious processes, this theory employs some conventional logic that rejects inconsistent propositions. Whilst it moves to the side of unconscious processes, if the symmetrization occurs extremely, then it will be inclined to a situation that includes everything.

In the modern non-classical logics, paraconsistent logics have gained some attention. Moreover, this attention is particularly directed toward the problems of measuring the degrees of inconsistency, incoherence in an ontology, and the reasoning (compare e.g. [24], [25], [26], [29], [30]). The common agreement of formulating the paraconsistent logics is to reject the ECSQ or to accept "the inconsistency without triviality".

We can say that the Matte Blanco's theory of mind formulated the symmetric-continuum of mind in a bi-logical perspective. Theoretically speaking, except two poles of the bi-logic framework, one pole adopts the consistent classical logic, and the other pole adopts the inconsistent logic that includes everything, the rest of symmetric-continuum has to be paraconsistent, since it admits some inconsistent propositions without the explosion. However, it is hard to say that the whole bi-logic framework is paraconsistent, since one pole of it rejects the explosion and the other pole accepts the explosion.

4.5 Discussion and Outlook

A certain sense of anti-mechanism stated that various human intelligence, expertise, and skills acquaintance which mainly rely on the unconscious instinct could not be taken in formal rules in this manner (compare e.g. [12], [13], [14], [15]). The core of this paper introduced the bi-logic framework that provides the theoretical underpinnings and gives psychoanalytic insights on the unconscious processes investigations. By some notions in mathematical logic, the formulation of basic rules in unconscious processes was given in Matte Blanco's theory of mind.

Concerning human intelligence investigations as our starting point, the unconscious processes rest on two main principles, namely an endorsement of the principle of symmetry and the principle of generalization, observed in the psychoanalysis studies have led to a systematic analysis of conceptual and logical problems in psychoanalysis that were previously not notified before Matte Blanco's works.

With the development of various logics in the modern logical society, several logics used in different domains today might be rendered as the options for the bi-logic framework. In the first place, various paraconsistent logical systems, including discussive logic, non-adjunctive systems, preservationism, adaptive logics, logics of formal inconsistency, many-valued logics, and relevant logics ([31]) can serve as the paraconsistent systems for studying the continuum of conscious and unconscious processes. Secondly, with the development of the classical logic and intuitionistic logic, the intuitionistic logic can serve as the replacement of classical logic as the conventional logic, since the intuitionistic logic is at least strong as the classical logic (compare e.g. [1], [18], and chapter 6 in [32]). Thirdly, by the result in modal logic that $S5$ is paraconsistent, so is first-order logic ([4]), we can extend the symmetric-continuum to cover the conventional logic that is assumed as classical logic by Matte Blanco's theory of mind. Fourthly, to formulate the bi-logic framework as context-dependent logic based on a sheaf-theoretical framework has been suggested by CJS Clarke ([9], [10], [11]).

The investigations about Matte Blanco's theory of mind and bi-logic framework have in particular been employed to define some general rules of unconscious processes, moreover to be some kind of formalization in an interdisciplinary manner. The rapid development of **institution theory** could suggest a picture that, we believe, will eventually supersede these notions about logic from a general logic viewpoint ([24], [27], [28]). For the intersection of theoretical computer science and unconscious studies, it will be a potential work to specify each individual logic of stratum in the whole bi-logic framework. In this way, the institution theory will serve as the new framework for the transition between conscious processes and unconscious processes that have already been discussed in this paper. For example, to institutionalize paraconsistent logics. A relative idea, which discusses the transition from nonconscious (superpositioned quantum information) to classical information has also been proposed by Stuart Hameroff ([22], p. 230).

In the perspective of philosophy of mind, a general challenge that will be addressed in the future is to find out the justification for the bi-logically structuring unconscious/conscious processes. A more fundamental debate should be as follows:

"Whether can the primary statement about "structuring human minds" be justified or not?"

Following abovementioned perspectives, people have argued that psychoanalysis, including Matte Blanco's theory of mind and bi-logic framework could have been viewed in the light of complex systems ([34], pp. 152-154).

A concrete future work concerning interfaces of complex systems includes an integration of bi-logic framework with the dynamic structures that has captured the mind structure, to incorporate a realm of mathematical chaos theory into the study of "unpredictability in living beings", and to add various applications to psychoanalysis. Here is such a voice:

> "[...] A system that is deterministic is one where, given a initial state, there is precisely one state that follows it at any one point in time [...] However, there are many natural systems that are indeterminate, especially living ones, where absolute precision of prediction is not possible, yet some limited or probabilistic prediction or expectation can be achieved." ([34], p. 154)

It gives a first handle on the important applications to bi-logic framework, of which some scientific areas across the complexity in the living systems, unpredictability, and the discovery of mathematically chaotic processes in the human intelligence investigations can be discussed.

Acknowledgements. Some relative works about this paper has been invited to present in the Agalma Foundation in Geneva. The author thanks the comments from Prof. Dr. François Ansermet, Dr. Mathieu Arminjon, Prof. Dr. Jean-Yves Béziau, Prof. Dr. Robert Hinshelwood, Prof. Dr. Pierre Magistretti. Prof. Dr. Daniele Mundici, and Prof. Dr. Hartly Slater.

References

1. Béziau, J.-Y.: Classical Negation Can Be Expressed by One of Its Halves. Logic Journal of the Interest Group in Pure and Applied Logics 7, 145–151 (1999)
2. Béziau, J.-Y.: Logica Universalis, 2nd edn. Birkhäuser, Basel (2007)
3. Béziau, J.-Y.: Paraconsistent Logic! Sorites 17, 17–26 (2006)
4. Béziau, J.-Y.: S5 is a Paraconsistent Logic and so is First-Order Classical Logic. Logical Investigations 9, 301–309 (2002)
5. Béziau, J.-Y.: What is Paraconsistent Logic? In: Batens, D. (ed.) Frontiers of Paraconsistent Logic, pp. 95–111. Research Studies Press, Baldock (2000)
6. Blanco, M.: Thinking, Feeling, and Being. Routledge, London (1988)
7. Blanco, M.: The Unconscious as Infinite Sets: An Essay in Bi-Logic. Duckworth, London (1975)
8. Bomford, R.: The Symmetry of God. Free Association Books (1999)
9. Clarke, C.J.S.: A New Quantum Theoretical Framework for Parapsychology. European Journal of Parapsychology 23(1), 3–30 (2008)
10. Clarke, C.J.S.: On the nature of bilogic: the work of Ignacio Matte Blanco (2006), `http://www.scispirit.com/matteblanco5web.html`

11. Clarke, C.J.S.: Both/And Thinking: Physics and Reality. In: Clarke, C.J.S. (ed.) Ways of Knowing: Science and Mysticism Today, pp. 143–158. Imprint Academic Publisher (2005)
12. Dreyfus, H.: What Computers Still Can't Do: A Critique of Artificial Reason. MIT Press, Cambridge (1992)
13. Dreyfus, H., Dreyfus, S.: Mind Over Machine: The Power of Human Intuition and Expertise in the Era of the Computer. Free Press, New York (1986)
14. Dreyfus, H.: What Computers Can't Do: The Limits of Artificial Intelligence (revised) (1979)
15. Dreyfus, H.: What Computers Can't Do: The Limits of Artificial Intelligence (1972)
16. Freud, S.: The Essentials of Psycho-Analysis (Vintage Classics), Vintage Classics, New edition (2005)
17. Freud, S.: The Unconscious (Penguin Modern Classics Translated Texts), Penguin Classics (2005)
18. Fu, T.-K., Kutz, O.: The Analysis and Synthesis of Logic Translation. In: Proceedings of the Twenty-Fifth International Florida Artificial Intelligence Research Society Conference: AI, Cognitive Semantics, Computational Linguistics and Logics (AICogSem), pp. 289–294. AAAI Press (2012)
19. Fu, T.-K.: A Formal Framework for the Unconscious Processes in Human Intelligence. In: Proceedings of the Twenty-Fifth International Florida Artificial Intelligence Research Society Conference: AI, Cognition and AI: Comparing Human Capability and Experience with Today's Computer Models, pp. 359–362. AAAI Press (2012)
20. Fu, T.-K.: A Study on Bi-logic (A Reflection on Wittgenstein). In: Munz, V.A., Puhl, K., Wang, J. (eds.) Language and World. Papers of the 32nd International Wittgenstein Symposium, Kirchberg am Wechsel/Lower Austria., vol. XVII, pp. 142–144. The Wittgenstein Archives at the University of Bergen (WAB)
21. Gabbay, D.: What is a Logical System? Oxford University Press (1994)
22. Hameroff, S.: Consciousness, Neurobiology and Quantum Mechanics: The Case for a Connection. In: Tuszynski, J. (ed.) The Emerging Physics of Consciousness, pp. 193–253. Springer (2007)
23. Jech, T.: Set Theory: The Third Millennium Edition, Revised and Expanded. Springer (2003)
24. Kutz, O., Mossakowski, T., Dominik, L.: Carnap, Goguen, and the Hyperontologies, Logica Universalis. Special Issue on "Is Logic Universal?" 4(2), 255–333 (2010)
25. Ma, Y., Hitzler, P.: Distance-based Measures of Inconsistency and Incoherency for Description Logics. In: Haarslev, V., Toman, D., Weddell, G. (eds.) Proceedings of the 23rd International Workshop on Description Logics (DL 2010), CEUR Workshop Proceedings, Waterloo, Canada, vol. 573, pp. 475–485 (2010)
26. Ma, Y., Hitzler, P., Lin, Z.: Algorithms for Paraconsistent Reasoning with OWL. In: Franconi, E., Kifer, M., May, W. (eds.) ESWC 2007. LNCS, vol. 4519, pp. 399–413. Springer, Heidelberg (2007)
27. Mossakowski, T., Goguen, J., Diaconescu, R., Tarlecki, A.: What is a Logic Translation? Logica Universeralis 3(1), 95–124 (2009)
28. Mossakowski, T., Goguen, J., Diaconescu, R., Tarlecki, A.: What is a Logic (revised version). In: Béziau, J.-Y. (ed.) Logica Universalis, pp. 111–133. Birkhäuser Verlag, Basel (2007)
29. Odintsov, S.P., Wansing, H.: Inconsistency-tolerant Description Logic. Part II: A tableau algorithm for CALCC. Journal of Applied Logic 6(3), 343–360 (2008)

30. Odintsov, S.P., Wansing, H.: Inconsistency-tolerant Description Logic. Motivation and Basic Systems. In: Hendricks, V., Malinowski, J. (eds.) Trends in Logic. 50 Years of Studia Logica, no. 21 in Trends in Logic, pp. 301–335. Kluwer Academic Publishers (2003)
31. Priest, G., Koji, T.: Paraconsistent Logic. In: Zalta, E.N. (ed.) The Stanford Encyclopedia of Philosophy (Summer 2009 edition), http://plato.stanford.edu/archives/sum2009/entries/logic-paraconsistent/
32. Priest, G.: An Introduction to Non-Classical Logic. Cambridge University Press (2001)
33. Priest, G.: Paraconsistency and Dialetheism. In: Gabbay, D., Woods, J. (eds.) Handbook of the History and Philosophy of Logic (2000)
34. Rayner, E.: Unconscious Logic: An Introduction to Matte Blanco's Bi-Logic and its Uses. Harvard University Press (1995)

Chapter 5
New Emergence as Supervenience Relieved of Problems

Eliška Květová

Abstract. Supervenience and emergence are remarkable notions of cognitive science, which notably influenced especially philosophy of mind in the twentieth century. Issue of supervenient or emergent relationships is complicated, but it is possible to observe a tendency to prefer emergence at the expense of the notion of supervenience from nineties. This paper aims to answer question why it is so and is based on the fact that the answer might be interesting for artificial intelligence, which often uses the term of emergence. These two notions have always been very close to each other. This paper introduces development and common history of the concepts as well as changes of their relationship. The goal of the contribution is not only to consider the history of supervenience and emergence, but also to find appropriate distance to introduce thesis explaining the current relationship between them. The thesis could be simply formulated as follows: New use of the concept of emergence can be understood as a continuation of the idea of supervenience deprived of its fundamental problems.

5.1 Introduction

Mind-body problem is one of the fundamental problems of cognitive sciences and its possible solutions were inspirations or foundations for number of streams in philosophy of mind as well as in artificial intelligence. This contribution aims to discuss relationship between two important notions which were strong or out-standing part of mind-body problem solutions' history. This chapter will deal with concepts of supervenience and emergence.

Ambition of this contribution is not to fully specify the nature of the relationship of the concepts. Such a goal would require a much deeper insight and detailed

Eliška Květová
Department of Philosophy, Faculty of Philosophy and Arts, University of West Bohemia, Sedláčkova 38, 306 14 Plzeň, Czech Republic
e-mail: eliska.kvetova@gmail.com

J. Kelemen et al. (Eds.): Beyond Artificial Intelligence, TIEI 4, pp. 75–84.
DOI: 10.1007/978-3-642-34422-0_5 © Springer-Verlag Berlin Heidelberg 2013

examination of the issue. As it turns out, the nature of the relationship to me largely depends on the level of description or better on the distance that we choose. I would like to point out some interesting aspects of the development of this relationship and would like to introduce a possible perspective on the role of the modern notion of emergence. The message or motto of the contribution is: *New use of the concept of emergence can be understood as a continuation of the idea of supervenience deprived of its fundamental problems.*

The initiative question is how it is possible that these originally closely related concepts got into the opposition according to many authors. It must be added that the history of these concepts is very long and rich, particularly when we take into account intuitive ideas of relationships which were later denoted as supervenient or emergent relationships. In this context we should go back to ancient Greece, I suppose. For this reflection I will focus on the context in which these terms appear in cognitive science, especially in philosophy of mind of the twentieth century.

The paper is based on an idea that the almost exclusive use of the concept of emergence in artificial intelligence is somehow cut off from its use and its relationship to the concept of supervenience in the philosophy of mind and other cognitive disciplines. One of the aims of the paper is to show that there is no reason for that. Emergence seems to be undetermined in many respects or differently determined in the field of artificial intelligence. In the next step, it is assumed that a detailed analysis, an insight to the history of this concept and its relationship to supervenience could be interesting, could prove useful for further development in the field of artificial intelligence. As Bedau expects according to [4], artificial life will play an important role in philosophical debates and discussions about notions like emergence, supervenience and other related issues like reduction, complexity or hierarchy. In both disciplines, in philosophy of mind as well as in artificial intelligence, mind and its properties, consciousness, emotions, wisdom or intelligence are the crucial issues and for both disciplines emergence seems to be promising concept nowadays. They will only differ in the ways how to treat these mental properties.[1]

If the cognitive science wants to take pride in being interdisciplinary, the mutual reflection and communication between its particular disciplines has to really be in presence, has to really work. This is one of the reasons why it is possible to take this paper to be a positive step in research and considerations of AI.

The structure of the chapter will be as follows. In the first part of the text the development of the relationship between supervenience and emergence will be described. Due to the chosen context that should enrich the view or the area of artificial intelligence, the notion of supervenience will be introduced first as the starting point of the interpretation. After this introduction, the development of the relationship between supervenience and emergence and thus the history of their mutual interactions will be approached. The issue of the differences and common features of supervenience and emergence will be discussed above all. The attention will be also paid to emergence as a concept that will be generally introduced and especially its position

[1] The possible way how to treat consciousness from the position of the theory of computation describes Kelemen in [14]. The philosophical point of view seems to be much more complicated or in different words worse seizable and more undetermined.

in artificial intelligence should be illuminated. Gradually I will get to the context of mind-body problem in which the concepts of supervenience and emergence are primarily considered in the twentieth century. The chapter will try to show the information about the development of their relationship as a guide for interpreting emergence as a new way to revive what once seemed attractive in supervenience in the context of mind-body problem.

5.2 Development of the Relationship of Emergence and Supervenience

Due to the way of interpreting the development of mutual relationship between supervenience and emergence which is presented in this chapter, reference to the notion of supervenience will be made first as it has been already notified (in spite of the fact that its history is connected with the notion of emergence which is even older).

Supervenience is usually introduced or presented (not defined) as a special type of relationship between properties.

> There is supervenience when and only when there cannot be a difference of some sort A (for example, mental) without a difference of some sort B (for example, physical). When there cannot be an A-difference without a B-difference, then but only then A-respects supervene on B-respects. [5]

> It is not possible that two things should be identical in respect of their lower-level properties without also being identical in respect of their upper-level properties. [10]

I deliberately avoid the term definition, because definition is something very problematic in this case and causes many difficulties to the notion. In the event of supervenience and emergence it is rather possible to meet with an approximation or definition-like and non-equivalent formulations. Because of the absence of complex specification of all aspects and characters attempts at definitions have never been accepted by the majority of the subject field.

But the simple and intuitive idea of supervenience was very attractive. The special type of relationship between two levels of properties is based on very intuitive idea that there is one kind of property or fact (supervenient property) that may only be present in virtue of the presence of some other kind of properties (subvenient base). New property appears and changes depending on the particular arrangement of underlying properties. It still maintains a degree of autonomy for supervenient properties.

The verb "supervene" derives from the Latin word "super", which means "on", "above", "additional", and from the Latin verb "venire" which means "to come". "Supervene" means "to come as an extraneous addition" according to [13], "coming or occurring as something additional, extraneous, or unexpected" according to [27].

Supervenience seemed to be very promising concept for many areas, especially for stated mind-body problem: mental properties are dependent on physical properties, but they have some autonomy. It was an attractive alternative to reductionist physicalism. It is possible to speak about boom of supervenience in cognitive science in the eighties of the twentieth century. This boom was connected with famous names as Jaegwon Kim, Terence Horgan, John Haugeland, David Kelogg Lewis, Simon Blackburn, Brian P. McLaughlin, Barry Leower etc. After some time there are many varieties of supervenience. But the initial optimism faded. Supervenience was largely stigmatised by criticism and many other difficulties (for example how to qualify dependency, covariance and non-reducibility of supervenient properties). Supervenience proved to be unable to answer the fundamental question of the nature of this concept, unable to resist the strong criticism. Supervenience with its problems is weakened or refused by many authors. On the other hand, emergence appears in philosophical (and not only philosophical) papers more and more frequently. From nineties there is a tendency to strictly distinguish between supervenience and emergence. It is interesting, because these two notions were used as synonyms in the twenties of twentieth century.

5.3 Development of the Relationship between the Concepts

It is indisputable that there is a close or even very close relationship between supervenience and emergence. Denotations of these words, terms, concepts have been changing for a very long time as well as their mutual position has been changing. Let me only briefly outline some important points of development of these concepts and their relationship.

McLaughlin in [18] starts the description of the use of terms emergence and supervenience by describing history of British emergentism. The next stop is therefore the tradition of British emergentism. John Stuart Mill is celebrated as father of this tradition. In chapter "Of the Composition of Causes" in [19] he distinguishes two modes of the conjoint action of causes—the mechanical and the chemical. And he also distinguishes between homopatic and heteropatic effect. In the chemical mode of the conjoint action of causes, the type of the effect of action of two or more types of causes is not the sum of the effects each of the causes—if we imagine that they acted alone or separately. He concretely shows an example of chemical compounds. If there is a combination of two substances, then this combination produces a third substance with properties different from properties of the two substances—separately and also of both of them taken together.

Other continued in Mill's consideration of the distinction between heteropatic and homopatic effect—for instance Alexander Bain or George Henry Lewis, who labeled the heteropatic effect as "emergent" in [17]. "Emergent" is an effect that is not only the sum of all effects of each of its causes if they acted separately. In Lewis' terminology heteropatic effects emerge from the causal factors that produce them.

Next name that should be mentioned in the context of British emergentism is Samuel Alexander, philosopher and theologian who describes in [2] emergent qualities or the emergence of a new quality from any level of existence. Biologist Llyod Morgan introduced the term supervenience into discussions of emergent evolution. Morgan in his [20] used words "supervene" and "emerge" as different stylistic variants of the same. He argues that through the process of evolution, new, unpredictable complex phenomena emerge. He connects idea of emergence with Darwinian evolution.

The last name that will be mentioned in terms of British emergentism is C. D. Broad. He describes an emergent property of a whole in [6] as property that could never be deduced from the knowledge of the properties of its components (taken separately and also taken in a variety of combinations), from the knowledge of arrangements of these properties nor from the knowledge of the proportion of these components' properties.

The next step in the development of the relationship between supervenience and emergence is Van Cleve's notion of an emergent property [24] which refers to supervenience. He employs the technical term of supervenience to define emergence.

5.4 Emergence in Artificial Intelligence

The notion of supervenience has been already introduced. To simply or in general determine emergence let me refer to Vision [25] who noted that common basis of emergentist theses was that some sorts of things "emerged" from an ontologically simpler foundation in ways defying rational expectation. If we look up the word "emergent" in the dictionary [26], we will find out that emergent (property or whatever) is appearing as something novel and in [26] it is directly connected with the emergent evolution which concerns the appearance of new and antecedently unpredictable qualities of being. These qualities could be for instance life or consciousness.

Modern concept of emergence occurs in many contexts, many areas. In this paper its position in cognitive science and especially in philosophy of mind and artificial intelligence should be stressed or accented. Emergence is connected with connectionist functionalism or with an interesting notion of the artificial life which I would like to mention.[2]

Artificial intelligence tries to model, tries to create systems capable of thinking, systems to which intelligence or other mental properties such as consciousness, emotions, etc. could be attributed. Very important is the fact mentioned for example by Havel [11] that any project of similar system will require a prior intention on one hand and an explicit design specification on the other hand. Similarly Cariani in [9] claims that the problem of emergence is primarily the problem of specification on one hand and creativity on the other hand or of closure and replicability vs. open-endedness and surprise. To particularize the examined phenomena an objective description of the process of the phenomena which should be modeled is needed.

[2] The next issue that could be discussed and analyzed in detail is the relationship between artificial intelligence and artificial life. Authors in [1] deal among other things with their relationship and compare artificial intelligence to an elder sister of artificial life. For the purposes of this paper, let us consider the younger field of artificial life as a part of artificial intelligence. The part that is based on the idea which was expressed for instance in [15], idea that the most important tool in the study of emergence is model building.

Taking into account that all properties which AI scientists try to model are not accessible to external observer, many problems appear. The attempt to specify thoughts, intelligence, emotions or consciousness seem to be insufficient to philosophers. AI definitions affect only certain features of the phenomena according to them. Mental properties must be something more, it is not possible to reduce them. This intuitive idea is again reflected by the concepts of supervenience and emergence. Moreover, even the field of artificial intelligence is akin to this view. Kelemen talks in [14] about an intuitive feeling that if we want to consider any robot consciousness, it will necessarily require something as attentiveness or emotionality. Emergence as a property of artificial systems seems to be very attractive. A set of grand challenges and problems in artificial life was formulated at the Seventh International Conference on Artificial Life [3]. Emergence or the demonstration of the emergence of intelligence and mind in an artificial living systems is in [1] mentioned as one of the open problems in artificial life.

As well as specifications for the project of any artificial phenomena is required, further specification of emergence is needed. This need leads in artificial intelligence to some kind of determination of emergence—to a test of emergence in artificial life. This test was proposed in [21] and [22]. It takes into account two scientists attendants: a system designer and a system observer, and three basic conditions: design, observation and surprise. This related conditions describe the conditions for diagnosing emergence, determine what system displays "emergent behavior". Philosophers would again criticize the limitation of these criteria, on the other hand their approach struggle with a total indetermination of emergence. The fact that even in artificial intelligence (not only in philosophy of mind) it is possible to observe a certain tendency towards diversification and fragmentation of the concept of emergence could be also interesting.[3]

5.5 The Nature of the Relationship between Supervenience and Emergence

In the nineties the researched terms of supervenience and emergence can no longer be claimed to be synonyms. Or as it turns out, they could be in a certain distance, but they are not. The nature of their relationship is increasingly confused and a number of questions appears. Is there any relationship of subordination and superiority between them? Is it not too simplistic to determine supervenience as term that could be employed to explicate the notion of emergence as McLaughlin in [18], as well as Van Cleve in [24], or Vision in [25] do it? Equally simplistic idea could be that emergence can be considered as a general term that refers to the appearance

[3] Havel in [11] distinguishes three meanings of the word emergence. Three views of emergence are also introduced in [4]. For this paper especially the second view is very interesting, because it refers to supervenience and supervenient properties which are irreducible. Also in [9] it is possible to find three "emergences": formally based computational emergence, physically based thermodynamic emergence and functionally based emergence relative to a model.

of "new properties" and supervenience gives stronger framework to emergence, because it indicates directly to the relationship between these "new" or supervenient properties and properties that could be called basic or subvenient.

It was stated that from nineties emergence appears more frequently in papers, while supervenience unable to successfully respond criticism loses its attractiveness. Attempt to define the simple idea of supervenience fails and this concept is gradually fragmenting into many types and categories or varieties of supervenience which could provide a more accurate definition (weak, strong, global, local etc.[4]). But because of this fragmentation supervenience loses its simplicity, clarity and appeal. Supervenience has been faced with many problems, the notion seemed to be somehow undetermined in many aspects. Supervenience fails to answer the basic question: What is it? What does it consist in? Why to believe that some properties supervene on another (base) properties? How can be explained that e.g. mental properties supervene on physical properties? Therefore the concept of supervenience had to recede into the background. Even supporters and proponents of this concept could not adequately answer questions about supervenience and provide a solid foundations for this notion. Their attempt to better specify the notion causes fragmentation, which unfortunately does not solve its problems, but brings a number of other problems and questions. On the other hand the idea of this kind of relationship remained to be an interesting possibility and inspiration.

Often discussed question in connection with mind-body problem is the question of ontological status. From my point of view, it is possible to speak about the ontological status of properties or phenomenon like mental state, emotion or consciousness for instance. But in the case of supervenience or emergence it is only possible to speak about the relationship between properties. The ontological status of supervenient or emergent property could be considered but in the case of supervenience or emergence as relationship between so-called upper and lower levels properties it does not seem to be appropriate.

There are tendencies to prefer emergence (during the BAI conference presentations it was possible to hear the term "emergence" or "emerge" several times, but no one used the term "supervenience"), or to prefer emergence and refuse supervenience. There is no need to be so radical as Humphreys in choosing title for his paper [12] seems to be, but it is necessary to highlight the fact that there is obvious tendency to advert to their differences, to distinguish between supervenience and emergence, e.g. in [8], [25], [18] or [23]. They all do it to fit these notions into their systems or theories.[5] In this context next question appears: In what do they differ in fact? Or how much do they differ? The answer leads us to the thesis of the paper.

---

[4] More about the fragmentation and about varieties of supervenience and their definitions in [16].

[5] The systems of these and other authors have different definitions of supervenience and emergence. Their mutual position is dependent on these definitions which do not correspond. Only one feature is common—stressing the fact that supervenience and emergence are not the same.

5.6 New Emergence as Continuation of Supervenience

I do not want to refute the views promoting differentiation and prioritization of emergence—they absolutely make sense when we consider problems and fragmentation of the notion of supervenience. In this sense new emergence tries to define or redefine the boundaries, to distinguish itself from problematic supervenience. My position is based on the opinion that it also makes good sense not to shatter the basic idea, to stay at a general level, to consider only the basic idea of these concepts—the prize that have to be paid is the absence of precise definition and the advantage on the other hand is that the core idea is not marked by stated problems and criticism. Both notions, supervenience as well as emergence, have problems with clear definition, but they share the same core idea. The thesis of modern emergence shows a striking resemblance with core idea of supervenience. There are many common features that are connected with not entirely successful supervenience as well as with modern emergence, features that could be labeled as heart of concepts of supervenience and emergence, that support my statement, that there is no big difference between core idea of supervenience and core idea of emergence.

First, it is possible to point out for example something that Vision [25] calls *a layered conception of reality*. Both notions share a hierarchical picture of the world, in both cases we speak about relationship between two levels of properties.[6]

Second, supervenience as well as emergence are so-called *topic neutral concepts*. The relationship of supervenience or emergence can be attributed to varied types of properties or facts. That is why they are applicable in many areas and we can read or hear about supervenience or emergence in philosophy of mind, chemistry, biology, sociology, ethics, aesthetics, artificial intelligence etc.

Third, the dictionary entries which have been already mentioned above should be reminded. All of them show that supervenient as well as emergent property or fact *appears as something additional, unexpected or novel.*

The next characteristic adverts to the fact that supervenience as well as emergence appears in dependence on *certain arrangement of basic properties* or in different words appears on a certain level of complexity.

The following feature concerns the fact that dependency relationship of supervenience or emergence should maintain *certain autonomy of supervenient or emergent property*. Both concepts share the idea that properties like mental state or consciousness are dependent on physical states, they are determined by them, but they have its own quality that cannot be reduced to physical state.

[6] According to Havel [11] it is not fully proper to talk about "levels" in the context of mental phenomena. In his opinion it is not possible to treat mental terms as something that should be confined to a certain "level", "domain" or "subject area". He admits the importance of inter-level interaction on distance, but doubts about embedding mental levels into the scalar hierarchy of functional levels of the brain. Nevertheless from my point of view, there is obvious and recurring tendency to understand mental properties as something that differs from the base, physical properties, as something more. In this context it is not possible to avoid the hierarchical ordering of reality, to avoid the idea of levels of reality.

The last characteristic I would like to mention is the possibility of *multiple realization* or *variable realization*, which concerns situation where two entities can be identical in terms of upper-level properties, but they differ in lower-level properties. Accordingly, the identical supervenient or emergent properties can supervene on or emerge from different base properties. This idea was very inviting for many streams in philosophy of mind, it also was important for artificial intelligence, because it anticipates that mental property, or mental state can be implemented not only by different physical properties of living organism, but can be implemented by different physical properties of any kind. What's more, the act of implementation should be avoided in this case. Mental properties can emerge from complex of physical properties of any kind. As professor Kelemen reminded from [7] in his speech for BAI conference, stuff like thought, consciousness or wisdom will not need to be programmed in, they will emerge.

If an intuitive idea that connects both concepts and all of the above are considered, the main thesis of the paper can be formulated: New use of the concept of emergence can be understood as a continuation of the idea of supervenience deprived of its fundamental problems. Emergence is an attempt to keep the good and attractive things about supervenience and distance from problems that this notion was not able to cope with. We put aside the fact how successful this second attempt was. The notion of emergence in artificial intelligence is largely connected with biological context and with the notion of evolution. That is why the fact, that the stated interpretation of new emergence as next stage of supervenience stopped by its problems makes sense in the context of philosophy of mind and of the issue of mind-body problem, has to be repeated.

5.7 Conclusions

The view of the development of concepts of supervenience and emergence can provide interesting conclusions or at least interesting realization or awareness of the interrelationships between philosophy of mind and artificial intelligence. Supervenience attracted much attention (not only in philosophy), but it failed to succeed. Time of emergence, of the concept which is artificial intelligence more familiar with, came. Two close concepts (supervenience and emergence) had to be distributed in order that the common idea can develop. Exploration of the history of these concepts can illuminate some aspects of their development, can answer asked questions, may allow a new interpretation, which is rather than big discovery a proof that shows how strongly the relationship which is in our discursus known as supervenient or emergent is deep-seated in our thinking or concepts, new emergence is just another stage of development of the concept of a hierarchical relationship between properties, a relationship that preserves some basic, subvenient properties which are arranged in a certain way. In this light it is possible to say that new use of the concept of emergence can be understood as a continuation of the idea of supervenience deprived of its fundamental problems.

References

1. Adami, C., Bedau, M.A., Green, D.G., Ikegami, T., Kaneko, K., McCaskill, J.S., Packard, N.H., Rasmussen, S., Ray, T.S.: Open Problems in Artificial Life. Artificial Life 6(4), 363–376 (2000)
2. Alexander, S.: Space, Time, and Deity. Macmillan & Co. (1920)
3. Bedau, M.A., McCaskill, J.S., Packard, N.H., Rasmussen, S. (eds.): Artificial Life VII: Proceedings of the Seventh International Conference on Artificial Life (Complex Adaptive Systems). MIT Press (2000)
4. Bedau, M.A.: Artificial Life. Blackwell Guide to the Philosophy of Computing and Information, pp. 197–211. Blackwell Publishing (2003)
5. Borchert, D.M. (ed.): Encyclopedia Of Philosophy. Macmillan (2005)
6. Broad, C.D.: The Mind and its Place in Nature. Kegan Paul (1925)
7. Brooks, R.A.: Cambrian Intelligence. MIT Press (1999)
8. Butterfield, J.: Emergence, Reduction and Supervenience: A Varied Landscape. Foundations of Physics 41(6), 920–959 (2011)
9. Cariani, P.: Emergence and Artificial Life. Artificial Life II—SFI Studies in the Sciences of Complexity, pp. 775–797. Addison-Wesley (1991)
10. Craig, E. (ed.): Routledge Encyclopedia of Philosophy. Routledge (1998)
11. Havel, I.M.: Artificial Thought and Emergent Mind. In: 13th International Joint Conference on Artifical Intelligence (IJCAI 1993), Chambéry, France, August 28-September 3, pp. 758–766 (1993)
12. Humphreys, P.: Emergence, Not Supervenience. Philosophy of Science 64(4), 337–345 (1997)
13. Johnson, S.: A Dictionary of the English Language. Richard Bentley (1755)
14. Kelemen, J.: A Short Note on Emergence of Computational Robot Consciousness. Acta Polytechnica Hungarica 5(4), 5–13 (2008)
15. Kilicay, N.H., Dagli, C.H.: Emergence and Artificial Life. In: IEEE International Engineering Management Conference (IEMC 2003), Albany, NY, USA, November 02-04, pp. 580–584 (2003)
16. Kim, J.: Supervenience and Mind: Selected Philosophical Essays. Cambridge University Press (1993)
17. Lewis, G.H.: Problems of Life and Mind. Houghton, Osgood and Company (1879)
18. McLaughlin, B.P.: Emergence and Supervenience. Intellectica 25(2), 25–43 (1997)
19. Mill, J.S.: A System of Logic, Ratiocinative and Inductive: Being a Connected View of the Principles of Evidence and the Methods of Scientific Investigation. Harper and Brothers Publishers (1859)
20. Morgan, C.L.: Emergent Evolution. Henry Holt and Co. (1923)
21. Ronald, E.M.A., Sipper, M., Capcarrère, M.S.: Testing for Emergence in Artificial Life. In: 5th European Conference on Advances in Artificial Life, ECAL 1999, pp. 13–20 (September 17, 1999)
22. Ronald, E.M.A., Sipper, M., Capcarrère, M.S.: Design, observation, surprise! A test of emergence. Artificial Life 5(3), 225–239 (1999)
23. Rueger, A.: Robust Supervenience and Emergence. Philosophy of Science 67(3), 466–489 (2000)
24. van Cleve, J.: Mind-Dust or Magic? Panpsychism versus Emergence. Philosophical Perspectives: Action Theory and Philosophy of Mind 4, 215–226 (1990)
25. Vision, G.: Re-Emergence: Locating Conscious Properties in a Material World. MIT Press (2011)
26. Webster's Third New International Dictionary, Unabridged. Merriam-Webster (1961)
27. Webster's New International Dictionary. Merriam-Webster (1986)

Chapter 6
Beyond Knowledge Systems

Jozef Kelemen

Abstract. The central subject of the present paper is to analyze the step-by-step process leading from data-based to knowledge-based systems, and to show the relationship between knowledge systems and the structure and roles of human wisdom. In order to do this, we formulate our main question: What can be seen when focusing beyond the knowledge systems? We also present one of the acceptable answers.

6.1 Introduction

The society which emphasizes knowledge in its own future development builds this future on the base of its own specific history. The specificity of this history is rooted in several traditions—legends, myths, narratives—which gave the humankind of the West a certainty to confide in the knowledge-based understanding of the world, and the rule-based behavior in it, in the past (and in certain extent they have provided it up to now). The rationalist philosophy of the Enlightenment was a spectacular reset of this confidence. The just emerging knowledge society will be perhaps the one which will be rooted on many successes and tragic failures of the previous one. Where are we now, and what are we able to expect in the future?

During the past few decades Artificial Intelligence (AI) and Cognitive Science as closely related disciplines brought numerous applicable results. One of them, and perhaps the most important, is a number of knowledge-based systems. *Knowledge-based* (or shortly *knowledge*) *systems* are those which include a *representation* of human knowledge, partly based on a human expertise, and also *inferring procedures* for the use of represented knowledge in order to *solve problems* the solving of which usually require knowledge and skills of human experts.

Jozef Kelemen
Research Institute of the IT4Innovations Centre of Excellence
and Institute of Computer Science, Silesian University, 746 01 Opava, Czech Republic
School of Management, 851 04 Bratislava, Slovakia
e-mail: kelemen@fpf.slu.cz

J. Kelemen et al. (Eds.): Beyond Artificial Intelligence, TIEI 4, pp. 85–94.
DOI: 10.1007/978-3-642-34422-0_6 © Springer-Verlag Berlin Heidelberg 2013

In connection with the success and the large applicability of knowledge systems, a question arises concerning the boundaries of their applicability as well as their impact on the life of human individuals and human society in the near future. These questions are evoked also by the realization of the relatively important changes at the level of education of human beings in the present time, especially by the remarkable decrease of the level of education. To be more precise, the specialists record not only the decrease in the level of professional preparedness of the young people in their particular disciplines, in their domain of expertise, but also in their general preparedness for participation in formulation and solution of more general problems, those which require *human wisdom*.

The central subject of this contribution is to analyze the step-by-step process leading from data-based to knowledge-based systems, and to demonstrate how knowledge systems are related to the structure and roles of human wisdom. In order to do this we formulate our main question: What will we see by focusing beyond the knowledge systems?

6.2 From Data to Knowledge—Role of the Context

First of all, let us emphasize an important role of the context in the development of using computers as primarily data processing devices.

The story started tens of years ago when the first computers have been constructed as suitable technical devices for *computing with numbers*, as machines for "number crunching". The input data interpreted as numbers of different formats have been "crunched" by suitable programs and the output data were interpreted again as numbers. In these cases, the context has been defined by the computer programs. The similar situation, but in the more explicit form appeared with the data of more general meaning and led to the development of special systems of computer programs called *databases*.

Putting an item into the database means to establish the relations of it to other data, i.e. to define a context in which the implanted datum will be interpreted by the given particular database system. In other words, by getting a datum and defining its context we receive a particular piece of information. The datum `65` as the string of symbols has in fact no meaning. But it may be, for instance, put into the context of the data `John` and `Age` (having predefined meanings), and we receive the information that `John is 65 years old`. If we change the context replacing the age by weight, we receive the other information: `John weights 65 kg`. We can conclude that the role of context in database systems appears to be crucial in the process of receiving information from data. The context provides the base for generating the results that we expect from the database systems, i.e. a new data. In the case of our example it is the issue of the answer to the question `How old is John?` for instance.

Information may be stored in computer memories. It is the basic idea of *information processing systems*. When we define the appropriate informational context we are able to generate the new information via computers. In case of our previous example, we can receive the new information that `John is slim` from the information `John is 65 years old` and the information `John weights 65 kg`.

However, the construction

```
IF ((John is 65 years old) AND (John weights 65 kg))
THEN (John is slim)
```

becomes more complex information now, it turns out to be an entity of a new specific category in the frame of our understanding. The entities of this category are usually called in the AI literature as *pieces of knowledge*. So that the pieces of knowledge implicitly present in the information systems enable us to receive answers to our questions in the form of new information.

6.3 The Structure of Knowledge

During the past few decades the "good old fashioned artificial intelligence" in the form presented e.g. in [13] provided several specific techniques for representing pieces of knowledge. Let us mention them briefly according to [5] (Chapter 4). The characterization of approaches consists in derivation of the basic representational formats of any *piece of knowledge* from three basic representational attributes. The specification of pieces of knowledge relates every piece of knowledge to the attribute of *declarability*, i.e. with the property that any knowledge can be declared or expressed in symbol structures. It is the base for any *symbolic representation* of knowledge, for instance in computer memory structures. Another attribute of pieces of knowledge is their *associability*— it is the ability to be associatively interrelated with some other such pieces in larger networks of concepts in order to characterize the complexity of its real-world aspects, and the semantics of the pieces in the context of related ones. The terms *associative* or *semantic nets* are usually used for denoting the resulting structures in artificial intelligence. The attribute of *procedurality* of pieces of knowledge refers to the possibility of manipulating the pieces of knowledge. Such manipulations may transfer them into new contexts or make possible the effective use of knowledge in different particular problem solving processes. Because each piece of knowledge has each of the three just mentioned attributes—it is declared in certain formalism, it has associative links with other pieces of knowledge, and it has its own procedural part prescribing how to use it—the emphasis put to each one of the above listed three attributes led to the development of more or less specific *knowledge representation schemes* which have positive but also negative properties during the history of artificial intelligence.

The effort to integrate the positive sides of all the just mentioned representation schemes, as well as to integrate them into a representational scheme of some other aspects of knowledge like *uncertainties* or *default values* etc., led during the development of Artificial Intelligence to different variations of schemes more or less similar to, but in basic principles almost identical with, the *frame representation scheme* as proposed in [6]. The resulting situation is depicted in Fig. 6.1.

Knowledge systems are characterized e.g. in [11, p. 312] as a term referring "... to a computer system that represents and uses knowledge to carry out tasks. This term

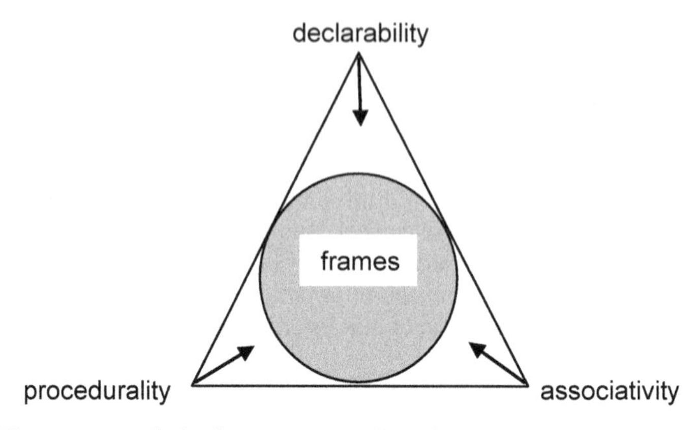

Fig. 6.1 The ways towards the frame representation scheme

focuses attention on the knowledge that the system carry, rather than on the question whether or not such knowledge constitutes expertise. The term domain refers to a body of knowledge." In this aspect knowledge systems slightly differ from the traditional expert systems focused more directly on the given domain of expertise. Knowledge systems contain pieces of knowledge mutually interrelated in a context of a given problem domain, or in the case of expert systems in a domain of given expertise. In this contextualized form the set of pieces of knowledge (the knowledge base of the knowledge system) create a coherent body of knowledge on the given domain.

Let us analyze the situation with the frame representation scheme in more details. We have mentioned that this scheme enlarged the representational possibilities of the previous schemes by the possibility to represent default values as contents of frame slots. Moreover, it makes possible to connect different frames into the frame systems and in consequences also a certain kind of the societies of knowledge going beyond the traditional associative connections between the pieces of knowledge towards the complex knowledge of the "societies of mind" [7]. This step is then followed by the introduction of the notion of the "emotion machine" [8] which introduces the study of machine emotions into the framework of the frame representation scheme. From this perspective, the Fig. 6.1 can be replaced by the Fig 6.2.

6.4 What Is Coming Now?

Now we are ready to return to a fairly urgent problem of the postmodern philosophers such as Richard Rorty, Jacques Derrida or Jean-François Lyotard. These philosophers look for answers to the question where can we today—in the world of cognitive pluralism, where freedom of an individual really has a high value and cultural differences and tolerance take place—plant a certain system of equality and where can we find a political and legal system, which we could subordinate to

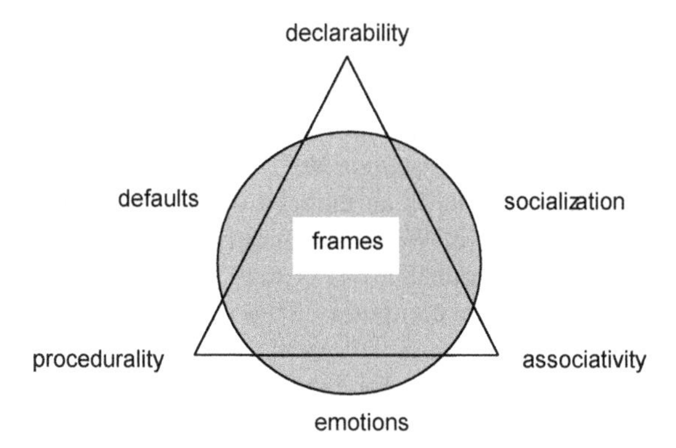

Fig. 6.2 The real contribution of the frame representation scheme into the understanding of the real structure of knowledge

ourselves. How can we reach an unforced agreement in these matters? This is a topic which is being discussed quite helplessly in philosophy.

What is coming? A society, in which there will be more space for individualization due to the new technology putting gradually its mark on the culture. Let us realize that if somebody actively stands up against the culture industry influenced by technology today, he should not see its roots in the first television broadcast, for instance, but rather in the invention of the printing press and in the ideology of people's equality and sovereignty, as puts it Umberto Eco [2]. When a recent intellectual finds him face-to-face with modern computer technology, he probably does not doubt the usefulness of the creation of conditions for general literacy or the creation and usefulness of a social structure, whose foundations could also be built thanks to the spread of literacy and therefore thanks to Guttenberg's—in principal technical—invention, and which determined the nature of a whole exceptionally important era in the development of our civilization. It seems to us more likely that something similar to the feelings of the former British proletarian survives—the fear that he would no longer be indispensable. However, every intellectual must himself overcome this barrier of the fear of estrangement.

The machine started to integrate into our society as soon as it was created and we started to use it. Reflexive interaction began, within the framework of which the machine has been improved, but at the same time, under the influence of the world it created, its creator has changed, too. The conditions of his life have changed his desires, hopes and the aim of his effort to create new machines, he has refined his vigilance, etc. It is enough to realize what influence the automobile has had on our behavior, our urban concepts, transport possibilities, legislation, etc. But to predict anything about the impacts of reflexive interaction is extraordinarily difficult.

If a reflexive interaction can change, writes George Soros influenced by Popper's concept of sciences and his own experience in analyzing developments on financial markets, *both the participant's thinking and the actual state of affairs, timeless generalization cannot be tested. What happened once does not necessarily recur when you repeat the experiment and the whole beautiful structure collapse. No wonder! Underlying the model is the unspoken assumption of a deterministic universe. If phenomena did not obey timelessly valid universal laws, how could those laws be used to produce predictions and explanations?* [10, p. 216].

What we now have before us is an embryo of the post-modern machine. But this is not a vision of the future! We must not only think of supercomputer superintelligence (cf. Ivan Havel's contribution in Chap. 1 of the present volume), skillful robots or a highly personalized computer of the future from the pages of science fiction.

What we now have on our mind is for example the cardio-stimulator. Probably in the future it will be more improved. Maybe it will be linked to a healthcare center, where it will send information about the state of the body, under the skin of which it has been implanted.

It will receive messages from the center on what to do in a given situation. It will be something more than just an electrical stimulator of muscle activity. It will maintain the optimum balance of different hormones, minerals, vitamins and who knows what else in the host organism and thereby protect it from stress, over-excitement, microorganisms and virus attacks, from the use of addictive chemicals and so on. Maybe we will not even carry cell phones in pockets and handbags. Maybe they will also be implanted somewhere under the skin. And they will function completely differently to that we use them today. Maybe a mere intention to announce something to somebody will be enough and he will know that we want to speak to him and if he has time and wants to, a connection will be made. There will be a town or a building that will follow your movements, receive your messages, and navigate you through its insides to get you where you will want to go. It will draw your attention to things that interest you, warn you of dangers... Maybe you will even be able to take a walk without leaving the comfort of your favorite chair, through strange visions and strange worlds of virtual realities.

We are just setting on a road which will soon bring us to the already mentioned very pragmatic understanding of the question which Jean Baudrillard formulated in a rhetorical and performative way—the question: *Am I a man or a machine?*

6.5 The Role of the Embodiment

The three attributes of the knowledge we have analyzed above lead us to finding the importance of the ability to socialize, to establish social and communication contacts, and to the importance of the emotions of the knowing beings including the artificial ones, e.g. robots or other sophisticated artificial agents. From that follows the possibility to draw the Fig. 6.3 in which the emergence of the consciousness is graphically sketched.

Consciousness will not need to be programmed in. They will emerge, stated R. Brooks [1, p. 185], one of the leading specialists of the present days artificial intelligence and advanced robotics research. The traditional and the widest informal

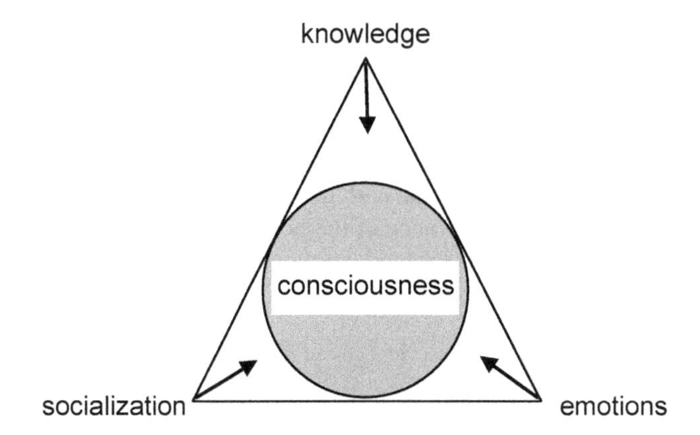

Fig. 6.3 Consciousness in the context of knowledge, social relations and emotions

definition of emergence has been formulated e.g. in [3, pp. 121–122]: Emergence is, according to him "*... a product of coupled, context-dependent interactions. Technically these interactions, and the resulting system, are nonlinear: The behavior of the overall system cannot be obtained by summing the behaviors of its constituent parts ... However, we can reduce the behavior of the whole to the lawful behavior of its parts, if we take nonlinear interactions into account*".

Intuitively, we feel that any robot consciousness necessarily requires attentiveness, sociability, and emotionality in the meaning we have mentioned in connection with the above sketched analysis of the frame representation scheme. Moreover, it requires also an attribute usually called the *private sense* (of the robot). To have the *private sense* means to have an ability of a given robot to consider itself as *another* robot identical with it, and to consider this type of "schizophrenia" in the work of other functions which characterize this robot. This type of recursion might be extremely complicated for expressing it in the frame of the traditional one-processor computational paradigm. It requires at least some suitable framework for dealing with behaviors that appear thanks to interrelations between individually autonomous robots.

The appearing situation insinuates the framework of considering a conscious robot as a system consisting of more independent but massively cooperating agents, i.e. in a form of multi-agent systems. The given robot's private sense is then perhaps an emerging product of interactions of its several functional agent-like modules, and might be also of a society of other robots which communicate in certain sense with the robot. Perhaps the conscious behavior of such a robot might be then described as a phenomenon, which emerges—in the above cited sense proposed in [3]—from interactions of traditionally computable behaviors of simpler constituting parts of it, and has the form of a hyper-computation. For more about that see e.g. [4].

However, thinking on robots requires the necessity to take into account also an inevitable structural and functional attribute of any real physically existing robot—the robotic *body*. Embodiment seems to be not only a consequence but also an antecedent part of any consciousness. It gives to a robotic being—through its sociability—also the attribute of its *"personal history"*. This statement provides a good base of executing a further step in the way of analyzing the perspectives of further qualitative development of knowledge based system. Similarly as we mentioned in connection with the consequences of the frame representational scheme for knowledge, we may do the similar with the concept of the body—we can look for the consequences of the appearance of the body (and the consciousness) in the context of knowledge, social contacts and emotions.

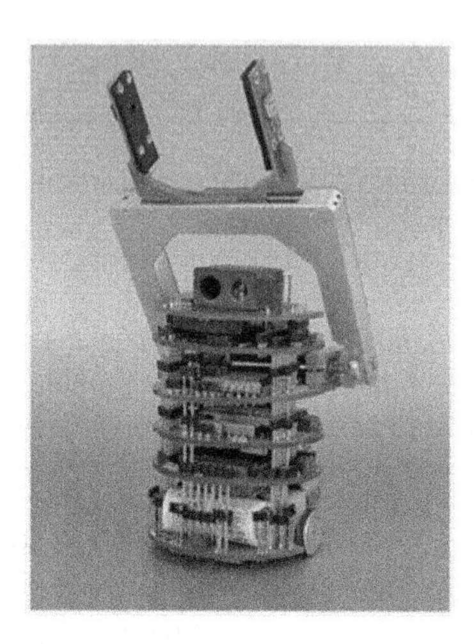

Fig. 6.4 An usual body of an usual (unconscious) experimental robotic platform

As we mentioned in connection with the consequences of the frame representational scheme for knowledge, we may do the similar with the concept of the body—we can look for the consequences of the appearance of the body (and the consciousness) in the context of knowledge, social contacts and emotions.

The result is sketched in the Fig. 6.5. which gives a position to the mysterious notion of wisdom.

6.6 Through Robotic Bodies towards the Computational Wisdom?

What kind of reasons can be considered as those drawing the boundary between the humankind and the machines of our time? One of the serious distinction consists in the existence of the human wisdom and non-existence of the machinized wisdom out of technical creations (at least up to now...).

However, consider now a realistic situation that we have lot of knowledge or expert systems at hand, so we are able to put into some contexts not only the knowledge pieces as in the case of building knowledge systems but also a number of

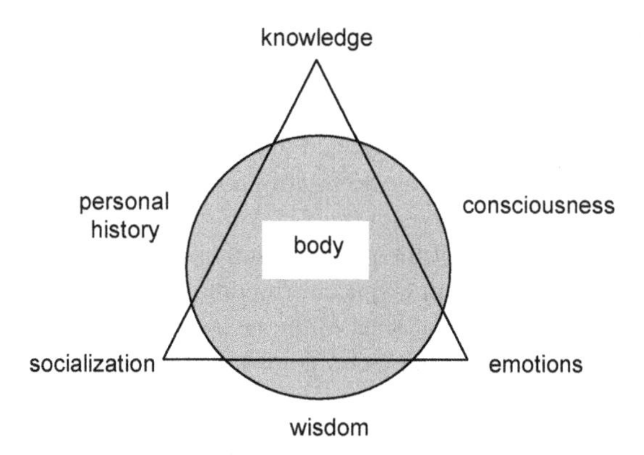

Fig. 6.5 The position of the wisdom?

different bodies of knowledge. What will we receive as the result of such a contextualization? In the case when the interconnected bodies of knowledge as the result will be on interrelated domains, we will receive a distributed or decentralized knowledge system. Approaching in such a way we will be able to organize more effective medical diagnostic knowledge systems with higher quality of solving problems in different interrelated domains of the medicine, for instance. What result we can image ourselves supposing the contextual interconnection for larger sets of knowledge systems in the case when the systems are not interconnected through their subjects at least in the first glance? May such kind of construction of decentralized or distributed knowledge systems result in any usable new knowledge system?

The usual answer is YES, and the usual expectation is that such kind of systems will reflect at least a part of an universal human intellectual capacity—a part of the human wisdom. *Wisdom* is—according to the Webster New World Dictionary [12, p. 1632]—"*... the power of judging rightly and following the soundest course of action, based on knowledge, experience, understanding, etc.*" It might be the above mentioned knowledge systems may contribute to better judging, to advice soundest course of actions, etc., namely if their bases of knowledge will be interrelated. There are several authorized questions in interest of present day research—see e.g. [9] for instance—especially in connection with the so called wisdom of crowds (the wisdom of certain kind of human societies). There are also authorized expectations such as that one on the emergent nature of the wisdom supposing that wisdom will not need to be programmed into our knowledge systems, that they will emerge similarly as the consciousness and thought will emerge in the case of robots [1, p. 185]. But what happens in the case if we consider another setting of the question? In other words, if we focus to the individuals, and if the context of the bodies of knowledge will be relatively narrow.

However, are there some boundaries for construction of some kind of computerized systems of human-like wisdom? We suppose that yes. Yes because of the human wisdom is constructed not only on the base of human common sense knowledge, not only on the base of human expert knowledge in different fields, but also on the base of human experiences, on human *common sense* as discussed by many authors—by Minsky in [7], for instance—during the history of AI. Generally, it is constructed on what it means to be a *human subject*, by his or her *personal history*, and the *history of other human beings*, and their past and their *culture* in the present or in the past, on his or her *experiences to have a human body* in the given place and time etc.

These facts, the existence of the personal authentic history, and the existence of the authentic body in the real time-space form, among others, the boundary, and we will be hardly in the position (at least in the near future) to be build a bridge over that boundary with our computing technologies.

Acknowledgements. Research of the author is supported by the project IT4Innovations Centre of Excellence, reg. no. CZ.1.05/1.1.00/02.0070, and sponsored by the Research and Development for Innovations Operational Program from the Structural Funds of the European Union, and from the state budget of the Czech Republic. The author thanks Petr Berka and Jana Horáková for numerous discussions concerning some of the topics discussed in this contribution, and the Gratex International Company for the continuous support of his research.

References

1. Brooks, R.A.: Cambrian Intelligence. MIT Press (1999)
2. Eco, U.: Apocalittici e integrati. Valentino Bompiani (1964)
3. Holland, J.H.: Emergence: From Chaos to Order. Oxford University Press (1998)
4. Kelemen, J.: May Embodiment Cause Hyper-Computation? In: Capcarrère, M.S., Freitas, A.A., Bentley, P.J., Johnson, C.G., Timmis, J. (eds.) ECAL 2005. LNCS (LNAI), vol. 3630, pp. 31–36. Springer, Heidelberg (2005)
5. Kelemen, J., et al.: Knowledge in Context. Iura Edition (2010)
6. Minsky, M.L.: A framework for representing knowledge. The Psychology of Computer Vision. McGraw-Hill (1975)
7. Minsky, M.L.: The Society of Mind. Simon & Schuster (1988)
8. Minsky, M.L.: The Emotion Machine: Commonsense Thinking, Artificial Intelligence, and the Future of the Human Mind. Simon & Schuster (2006)
9. Raisanen, T.: Five Experiments of Wisdom of Crowds as a Source of Knowledge. In: 6th International Workshop on Knowledge Management, Trenčín, Slovakia, October 21-22, pp. 154–160 (2011)
10. Soros, G.: Soros on Soros: Staying Ahead of the Curve. Wiley (1995)
11. Stefik, M.: Introduction to Knowledge Systems. Morgan Kaufmann (1995)
12. Websters New World Dictionary, Second College Edition. Prentice Hall (1980)
13. Winston, P.H.: Artificial Intelligence. Addison-Wesley (1977)

Part II
Nature-Inspired Models

It always strikes us when we realise that a particular nature-inspired model works so well. But is it really so surprising that it happens to be a nature-inspired model that works so well for a nature-inspired goal? Yes, it is. We do not stubbornly mimick birds anymore when we construct flying machines. So what about thinking machines? Will chaotic bio-inspired systems or membrane computing help us? Have we noticed that our experiments with evolutionary computing might be related to the selfish gene concept?

Chapter 7
Selfish Genes and Evolutionary Computation

Jan Zelinka

Abstract. This paper deals with the relation between the so-called selfish genes and evolutionary computing without wishing to immerse into the biological evolution theories. The main goal is to show how a selfish gene could appear and how it is possible to demonstrate the presence of a selfish gene. We also want to answer the question if and how can the selfish gene be beneficial in the evolutionary computing.

7.1 Introduction

The main topic of this paper is Evolutionary Computation (EC) and the relation between ECs and selfish genes (SGs). In this paper, we understand ECs narrowly as optimization methods where a population of temporary solutions is iteratively modified by means of stochastic operator selection and reproduction, while the result of a selection operator is determined or at least influenced by the value of a fitness function that constitutes some kind of a measure of the suitability of a solution. This paper has two goals:

1. To try to describe a way how a phenomenon which can be interpreted as a SG manifestation can emerge in ECs.
2. To try to present ways in which a SG manifestation can be beneficial.

Somehow to the contrary to the second goal presented above, the SG also poses some limitations to the particular EC although we do not always mention it explicitly.

The definition of most of the concepts used in this paper is often problematic or at least vague. Especially, the notion of "gene" belongs between controversial concepts. Nevertheless, we avoid thorough analysis of the utilized concepts (except

Jan Zelinka
Department of Cybernetics, Faculty of Applied Sciences, University of West Bohemia,
Univerzitní 8, 306 14 Plzeň, Czech Republic
e-mail: zelinka@kky.zcu.cz

J. Kelemen et al. (Eds.): Beyond Artificial Intelligence, TIEI 4, pp. 97–111.
DOI: 10.1007/978-3-642-34422-0_7 © Springer-Verlag Berlin Heidelberg 2013

for "gene") on purpose although we find the corresponding controversy interesting and useful.

In order to avoid misunderstandings, it is necessary to differentiate two standpoints from which the evolution can be viewed. That is, the standpoint of empirical science and the standpoint of engineering. The main difference between empirical science and engineering is an attitude towards relating real world objects to abstract models Empirical science usually modifies its abstract models whenever there is a discrepancy between them and the real world, whereas engineering attempts to modify the real objects to fit the abstract models. We suppose that the evolution theories are empirical sciences and ECs surely belong to engineering. So, in ECs a correspondence between observation in nature and model of evolution is not as essential as in evolution theories. However, feasibility of an artificial environment and a utility of the environment are essential. An engineer can observe nature and inspire in nature and the inspiration can be productive even if he later reveals that his observations were wrong. Theory of SG seems to be consistent, but we could object that certain observations falsify some evolution theories (and we think that Neo-Darwinism is falsifiable) or there could be some objections that this type of Darwinism is not suitable for nature and it is suitable only for some simple artificial systems. Now it should be clear that the last objection is not obviously an objection against ECs.

7.2 Evolution and Selfish Genes

This section is focused on interpretation of the SG concept and on hypothesis about the origin of selfishness. We present here our opinion about the role of reproduction in the origin of evolution and selfishness.

7.2.1 A Gene Definition

We presume that it is not necessary to find accurate formal definition of a gene because we are not going to identify any particular gene. But it is important to specify what type of objects genes are . There is a lot of definitions of gene and also there are controversies over these definitions in the literature. The most recent is the paper [1]. Richard Dawkins in [2] refers to the definition by C. G. Williams: "A gene is defined as any portion of chromosomal material that potentially lasts for enough generations to serve as a unit of natural selection."[2] In the same book Dawkins presents another definition based on the function of a gene which does not say from which material a gene is made: "I said that I preferred to think of the gene as the fundamental unit of natural selection, and therefore the fundamental unit of self-interest."[2] There is a lot of controversial aspects in these or similar definitions [3] but now we are going to focus on one single controversy. It is obvious that two different organisms which share the same genes do not share the very same portion of chromosomal material in the same way in which neighbors share the same elevator. So, a gene must be some kind of an abstract structure. Any evolutionist who seriously deals with memes [4] must obviously subscribe to this definition. We consider this notion of gene to be important if a concept of gene should be applied in ECs.

If an evolution should have the mentioned characteristics that make it an exceptional tool, reproduction or selection must be a stochastic process. Furthermore, an evolution is possible only if reproduction is not accurate. But then the concept of gene as a structure or information that survives the reproduction unaltered is not possible since in the framework of inaccurate reproduction the original gene ceases to exists and a new, somewhat different, gene comes up. So, a gene must be rather a class of structures or information. This might seem to be an arbitrary abstraction; nevertheless, this concept is not dissimilar to the concept of gene which we use in this paper.

We believe that the evolution theory has the same right to create abstract constructions as physics does (i.e., constructions like velocity or density.). On the other hand, we cannot allow the evolution theory to assert the existence of such objects or even let it assert that those phenomena are observable; in the same way as we can not deduce an indisputable existence of mock turtle from the indisputable existence of mock turtle soup. Even though we deny genes the property of being a directly observable phenomena, we do allow them to manifest some of their properties which then can be observable and thus falsifiable. Thus we rather refer to the manifestation of a gene instead of a gene itself in this paper.

7.2.2 Scattered Genes

The term Scattered Gene usually means something different in the literature than in this paper[1]. We changed the meaning as follows: A scattered gene is a gene which has different parts in different vehicles. Dawkins noticeably does not stand up for group-selection. In our opinion, it is because the subject of group-selection Dawkins cannot consider a gene. The idea that different substructures in two different molecules in different vehicles can be called as a gene is somehow controversial; however, since a gene is still an abstract object, we cannot find arguments against such concept in any observations. Let us consider that different parts of a gene can be not only in different vehicles but also in different vehicles in different times. Of course, this leads to a total gene concept deconstruction, even to the situation where rocks or sun could be parts of a gene. We are not going to advocate this idea a priori even in ECs. But a concept of gene determines what we can recognize as a manifestation of a SG. Thus, bellow we will mention these gene concepts and resultant types of manifestations.

[1] "Most clustering algorithms assign all points into clusters. However, in microarray experiments, we expect many genes to be unrelated to the biological processes that we are investigating and to show uncorrelated variations with any cluster of genes. These genes should not be assigned into any specific cluster, and are thus called scattered genes."[5]

7.2.3 Sources of Evolutions

Two main concepts in an evolution theory are reproduction and selection, where "selection" means selection of genes for further reproduction. Selection in evolution theories is classified in many ways. The most useful classification is classification according to causes of selections. There is a classical classification: natural selection and sexual selection [6]. However, this classification seems to be questionable.In the modern literature there are much more exact classifications based on the result of the selection used in a particular algorithm [7] but these classifications do not show causes of particular result of particular selection. Thus, we will use the term selection without further specification.

The amount of space devoted to discovery and description of selection in many monographs could lead to a belief that the selection is a single one decisive factor in the evolution. But this belief would be absolutely wrong, not only because there is no evolution without a reproduction. In the following text, we want to show that the real decisive factor is in fact a reproduction. Now it is important to answer this question: Can there be any type of evolution without a selection? Naturally, a selection cannot be canceled in nature but it can be canceled in an EC.

Let us imagine this algorithm: There are objects in the algorithm. Each object a has one single attribute: positive rational number a_T. Each object a periodically, with period a_T, creates randomly a new object b where

$$p(b_T) = \begin{cases} \frac{1}{2} & b_T = \frac{a_T}{2} \\ \frac{1}{2} & b_T = 2 \cdot a_T \\ 0 & \text{otherwise} \end{cases} . \tag{7.1}$$

The algorithm starts with one object a with $a_T = 1$ and the first reproduction happens in the time 0. There is no selection in this algorithm because there is no restriction for reproductions and no object cease to exist during the algorithm execution. There is only reproduction here. Although there are many different scenarios how the population can grow, only one scenario is probable: hyperbolic growth where the number leads to the singularity. We can reach this conclusion either by sheer consideration or by running a simulation experiment. Our simulations of the algorithm execution are shown in Fig. 7.1. So, there is an evolution with the values of the object attributes descending to zero, controled only by means of the reproduction mechanism. It is interesting that even in this very simple experiment evolution leads to the behavior which we are going to call selfish later in this paper.

We can extend the aforementioned algorithm by a fitness function, which will lead to an EC. The EC could be designed in the same way as the algorithm described above, only the object periods would depend on the values of the fitness function. Such an algorithm would certainly be impractical because it would need huge memory but it would nevertheless maximize the fitness function. We have tried a modification where fitness function was defined as

$$f(x) = \mathrm{e}^{-(x+2.5)^2} + \mathrm{e}^{-(x+1)^2} + 10\mathrm{e}^{-1000(x-1)^2} \tag{7.2}$$

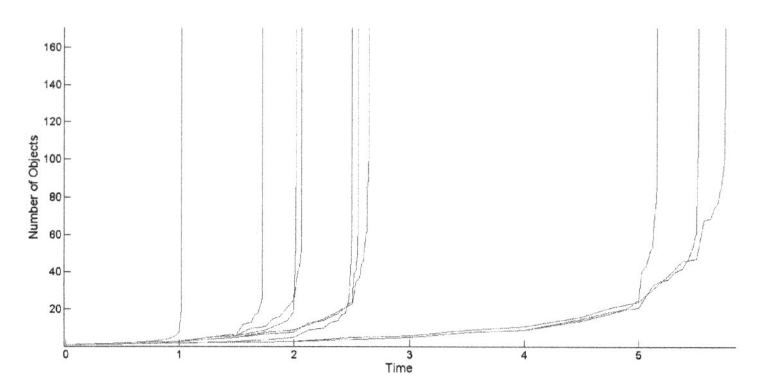

Fig. 7.1 Results of the algorithm described in Sect. 7.2.3. Each curve means a single simulation of the algorithm.

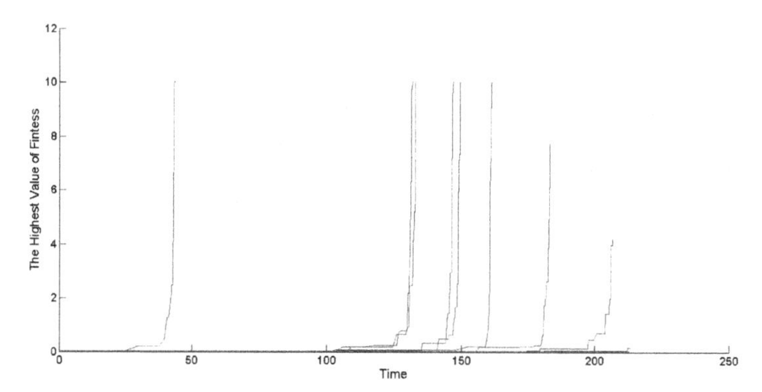

Fig. 7.2 Results of the modified algorithm described in Sect. 7.2.3. Each curve means a single simulation of the algorithm.

and a period was $a_T = \frac{1}{f(a_x)}$ where a_x is the second attribute of an object a. Some results are showed in Fig. 7.2.

Surely, the interpretation that a reproduction is an engine of evolution and a selection is only a catalytic is not the only possible one. Another interpretation is that there is not a sharp border line between reproduction and selection. But even this interpretation does not impede us to fulfill the first goal of this paper, i.e., to prove that the way of the selection in ECs does not make a SG manifestation impossible.

7.2.4 Evolutionarily Stable Strategy and Evolutionarily Stable Population

Usual concept in modern evolution theories is evolutionarily stable strategy (ESS) [8, 9]. Although the issues of ESSs are well known, our paper still the next short paragraph for a brief explanation because this concept is very important for all the following argumentation.[2] One might think that ESS definition makes sense only if there is an artificial criterion of optimality which measures something different than gene survival. But it would not be true because a difference between optimal strategy and ESS can be observed also in the case of the criterion which is defined as probability of gene survival. For example, it is a strategy of reproduction in a limited environment such as reproduction of some pathogenic organisms in an infected organism which is not immune from the disease. The optimal strategy is the strategy which leads to long life of the infected organism, i.e. the strategy numerus clausus. The optimal strategy is not evolutionarily stable because a gene leading to unscrupulous fast reproduction would have indisputable advantage and it would prevail over the restrained ones. We realize that such concept of self-interest might include an idea that the existence of a gene has intensity, which can be e.g. the number of vehicles, but rather something more complex.

Besides ESSs, we would like to speak about evolutionarily stable population (ESP). An ESP is a population where a change of strategy of one individual population member leads to the member's lower profit. Population where all members have ESS is obviously ESP. Another case is for example a population of flowers and pollinators with some particular ratio. Similarly, one can speak about evolutionarily stable cycles or even about something more general. Even Dawkins is apparently familiar with concept of ESP.[3] Concept of population consisting of "doves" and "hawks" is probably only simplification of mixed strategy concept which was necessary in a popular book. Nevertheless, even usage of this simplified example can be interpreted as willingness to admit the concept of ESP.

7.2.5 *Selfishness and a Selfish Gene Manifestations*

In this paper, the term selfishness always means striving for survival and the ignorance of any other aims, profits etc. Needless to say, selfishness in case of genes refers to the survival of genes, not their vehicles. A SG would emerge (and manifest itself) if evolution can influence the mode of reproduction and/or selection in some other way than by means of some deterministic mechanism.

Firstly, such manifestation is a difference between the optimal strategy and an ESS. Certainly, we do not mean a situation where an EC gets stuck in a local extreme (that would be indefensible arbitrariness). We mean the situations where the global extremes of a fitness functions are not ESS; i.e., although a population reaches a state where all members have the optimal strategy, the population leaves that state

[2] "An evolutionarily stable strategy or ESS is defined as a strategy which, if most members of a population adopt it, cannot be bettered by an alternative strategy."[2]

[3] "The way I have told the story it looks as if there will be a continuous oscillation in the population. Hawk genes will sweep to ascendancy; then, as a consequence of the hawk majority, dove genes will gain an advantage and increase in numbers until once again hawk genes start to prosper, and so on. However, it need not be an oscillation like this. There is a stable ratio of hawks to doves."[2]

and it stabilizes in some other state with lower rating. We leave aside the situations where no ESS exists. Already there is an evident condition of this manifestation, i.e. no members (not even the fittest ones) may survive permanently. We see this situation as a SG manifestation because there is a gene survival preferred to any other goal.

The second way of a SG manifestation is a frozen evolution. We do not mean the very same frozen evolution as it is mentioned in the monograph [3]. It would mean that even a scattered gene is a gene. We simply mean that inaccurate reproduction disappears. Of course, it must be added that a gene must be able to stop evolution in this way. Furthermore, the evolution must be frozen in a non-optimal state of a population. We see this as a SG manifestation because a gene, by preventing its incorrect reproduction, prevents its conversion and consequently its disappearance.

7.3 Evolutionary Computation

This section would first like to focus on the description of the EC and its role in optimization [10, 11]. First and foremost, we would like to specifically emphasize that we do not see neither the EC as an omnipotent optimizing method nor the SG [2, 12] as a powerful concept which can provide a simple holistic model of the world. Instead, we intend to present some clues that could lead us to more reserved but still positive view on at least the EC.

The first thesis, for which we will provide some arguments, is the assertion that ECs are approximations of the (generally unknown) optimal solution method. This seems to be entirely correct because ECs generally do not give the best possible solution. But let us consider a case when such a solution can be regarded as existing. Of course, there are some unsolvable tasks. For example there is no such thing as isopentahedron but something similar can by obtained by means of an EC. However, we do not mean only this situation. Let us deal with tasks which have a solution from this perspective but the solution is not analytical. No one surely can expect that some miraculously informed oracle would give us our solution but we can follow an indifferent mechanism which can do the steps which cannot be done by analysts and which are not always worse than analyst judgments. Just ECs with random numbers generators can arrange this. This makes ECs very extraordinary tool for the optimization problem. We descried the role of ECs in Artificial intelligence in Sect. 7.5.

7.4 Selfish Genes and Evolutionary Computation

Firstly, we will discuss a publication that deals with our topic of interest. Secondly, we will proceed to fulfil the declared goals of our paper. We will focus mostly on examples because any mathematical reasoning is relatively difficult for ECs.

7.4.1 State of the Art

In this section, we will deal with SG in EC and its reflection in related literature. First, we want to point out the fact that, even in extensive monographs, any connection between the selfish gene notion and the EC is usually totally disregarded or the selfish gene concept is just mentioned in passing without the real consequences for the EC algorithms. We believe that it is because most authors presume that the introduction of a fitness function totally eliminates selfish gene and its harmful effects. Fortunately, there are certain significant exceptions to this approach [13, 14, 15, 16, 17]. A short resume of the paper by Fulvio Corno, Matteo Sonza Reorda and Giovanni Squillero [13] dealing with SG concept within EC is presented in this section. The authors find the cause of SG absence in ECs in the assumption that a subject of a selection in ECs is an individual member of a population, whereas in evolution theory described by Dawkins the subjects of a selection are genes.[4] This somewhat confusing hypothesis leads the author to the conclusion that the main selection subject must be a gene.[5] So, in the algorithm proposed in [13] it holds true that

1. the algorithm does not explicitly use elitism,[6]
2. a "gene value" called "allele" fight for chance that it will be in a place called locus.[7]

Given those conditions, the SG should emerge. We did not find in [13] any account on if or how the emerging happened. But on the example of this paper we would like to show that the authors were right and that especially the first step is important and the second step is not so substantial for a SG manifestation. The main advantages are, according to the authors' opinion, better (more accurate) results and faster convergence.[8] We would like to show in this paper that a SG manifestation does not ensure better results and that these types of ECs can have other advantages.

7.4.2 Selfish Gene Manifestations in Examples and Experiments

Let us have a fitness function (7.2) and a standard EC algorithm A which process as follows:

1. A population P_0 consists of n ransom solutions.
2. For $t = 1, 2, \ldots, T$ repeat:

[4] "In a population, the important aspect is not the fitness of various individuals, since they are mortal, and their good qualities will be lost with their death. (...) Genes are selected by evolution on the basis of their ability to reproduce and spread in the population: the population itself can therefore be seen as a pool of genes."[13]

[5] "Algorithms, whose focus is on the fitness of genes, rather than of individuals."[13]

[6] "... a mechanism called *elitism* is used to preserve best individuals through generations, giving them a sort of unnatural longevity, or even immortality."[13]

[7] "In the SG, different alleles fight to be present in a specific locus."[13]

[8] "With this mechanism, alleles of the winner increase their selection probability, forming a positive feedback that drives a fast algorithmic convergence."[13]

a. Create a new population P'_t consisting of mutated members of population P_{t-1}. Each member of the population P_t is mutated $m > 1$ times.
b. $P'_t \leftarrow P'_t \cup P_{t-1}$.
c. Evaluate all members of P'_t by means a fitness function f and create a population P_t which consists of n best members of P'_t.

For this algorithm there is no described SG manifestation because if all members of a population are in the optimum, they will stay in the optimum and this is the only situation when this evolution becomes frozen.

Let us omit the step 2b in the algorithm A and create a new algorithm A'. And let us assume that the probability that a mutant is different from its original individual and even from any other member of the population is 1. So, each member (vehicle) survives one step of the algorithm A' with probability which equals to 0. Specifically, let us assume that a mutant x' is a random variable with normal (gaussian) probability density function $P_{mut}(x'|x) = N(x',x,\sigma^2)$ where x is the original member and $\sigma^2 > 0$ is a mutation parameter. Fig. 7.3 shows the fitness function defined in (7.2). One can see that the function has one global maximum G and two local maxima L_1 and L_2. Let us leave values of the fitness function for a vehicle x because it will "die" during one step of A' and let us focus on the mean of the fitness function of the mutant of the vehicle x. For this mean $F(x)$ holds $F(x) = \int_{\xi \in \Re} f(\xi) P_{mut}(\xi|x)$. (It would be more suitable to compute mean fitness of mutant of mutant of ... of mutant of the original vehicle x. But there are some technical difficulties.) Now the x^* with the highest value $F(x)$ is important. It obviously depends on the parameter σ^2. We estimated some points representing arguments of the maximum of the function $F(x)$ and we draw the points in Fig. 7.4. (The "wish bone" is in the figure because for some σ^2 we used an approximation where an argument of the maximum has two solutions.) So, one can see that for low σ^2 the x with the highest value of F is very close (it does not equal) to the optimum G. For higher σ^2, the graph goes over to regions which are close to local extremes and finally it leaves even this regions. Fig. 7.4 shows that for some σ^2 a gene which is not the optimal solution could be evolutionarily stable. We want to emphasize that σ^2 is a parameter of reproduction not selection.

However, this result does not mean that the same will be true for next generations. We performed several experiments to verify the hypothesis that the optimum is not evolutionarily stable. We fixed σ^2 as 0.28. Population P_0 consisted of the optimal solutions G. The result of one experiment is in Fig. 7.5. Each asterisk represents one particular member of the population. The evolutions in the other experiments were very similar. In Fig. 7.5, one can see that the population "escapes" from the optimal state to some more stable state. Thus, these experiments show that our hypothesis was correct, i.e. that there is the SG manifestation in this EC.

So, the result of the algorithm A' strongly depends on σ^2. Now, we add the parameter σ^2 into the genes. We fixed mutation operator as follows: $P'_{mut}(\xi, s^2|x,\sigma^2) = N(\xi,x,|\sigma^2|)N(s^2,\sigma^2,|\sigma^2|)$. One can see that ESSs are all strategies with $x \in \Re$ and $\sigma^2 = 0$. But the fact that a strategy is ESS does not always mean that an evolution heads towards the strategy. There are noticeable differences between those

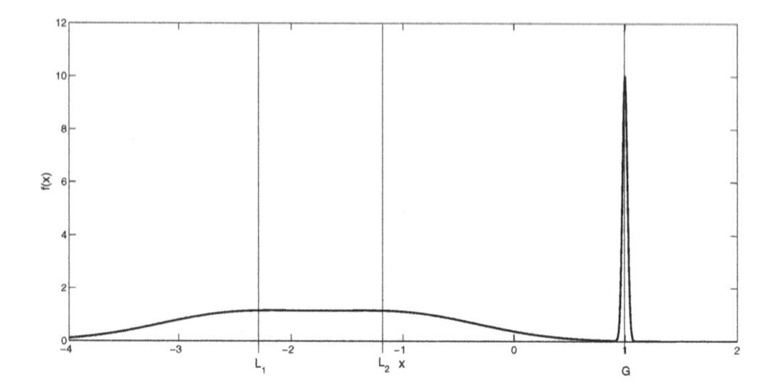

Fig. 7.3 The used fitness function. (See equation (7.2))

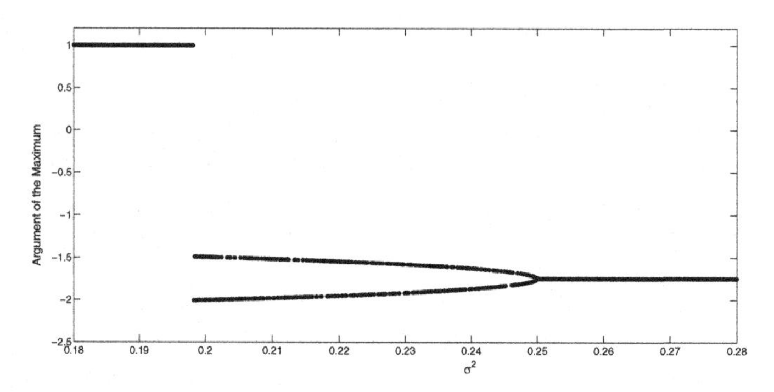

Fig. 7.4 Function F extremes for different values of σ^2

ESSs. Let us define an attractive ESS as an ESS for which exists a neighbourhood where there is no ESS with higher value of the fitness function. We believe that these strategies are points towards which the evolution heads (under some conditions which are in our opinion fulfilled). One can see that there are three such strategies: $(x = G, \sigma^2 = 0)$, $(x = L_1, \sigma^2 = 0)$ and $(x = L_2, \sigma^2 = 0)$. So, we avoided the first SG manifestation but we face the second manifestation, i.e. frozen evolution.

7.4.3 Advantages of Selfish Gene in Evolutionary Computation

The second goal of this paper is to try to describe a way how SG manifestations can be beneficial. There is no reasonable argument that would support the claim that the main advantage of some ECs with a demonstrable SG manifestation are better (more accurate, successful, etc.) results. The opposite may be true, especially when the manifestation is the difference between the optimal strategy and the ESS. Faster convergence would certainly be an advantage but we have no clue how to prove that it is the SGs that causes it.

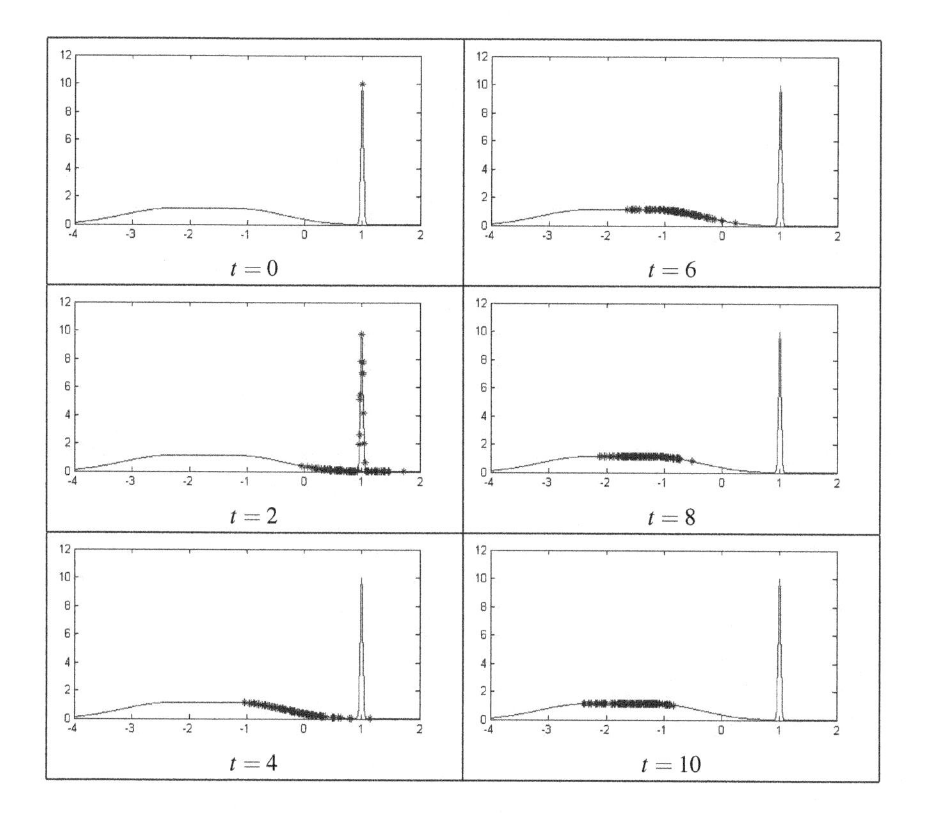

Fig. 7.5 The experiments results

We think that the important property of the solutions which are yielded by SGs could be robustness. Let us show this using the following example. We have a set $S = \{o_1, \ldots, o_{n_S}\} \subset \Re$. We want to maximize observation S likelihood, i.e. we want to find Θ which maximizes

$$L(S|\Theta) = \log P(S|\Theta) = \sum_{o \in S} \log(P(o|\Theta)), \; P(o|\Theta) = \sum_{i=1}^{n} \gamma_i N(o, \mu_i, \sigma_i^2), \qquad (7.3)$$

where $\Theta = \left[\gamma_1, \mu_1, \sigma_1^2, \ldots, \gamma_n, \mu_n, \sigma_n^2\right]^{\mathrm{T}}$. It can be very easily proved that there is no global extreme for $n > 1$ and there are only local extremes. But this function has its poles in improper points [18] (see Fig. 7.6) where solutions which are close to the poles are collapsing solutions and they represents extreme overtraining. The thin peak in Fig. 7.6 represents the collapsing solutions. Sensible solution is a locale extreme. There is a lot of ways how this collapse can be prevented. (See [18] page 434.) We believe that algorithms which are similar to the algorithm A' (see Sect. 7.4.2) represent such a suitable approach. We see the reason why the solution G is not an ESS (see Fig. 7.3) in the insufficient width of the "hill" whose top is

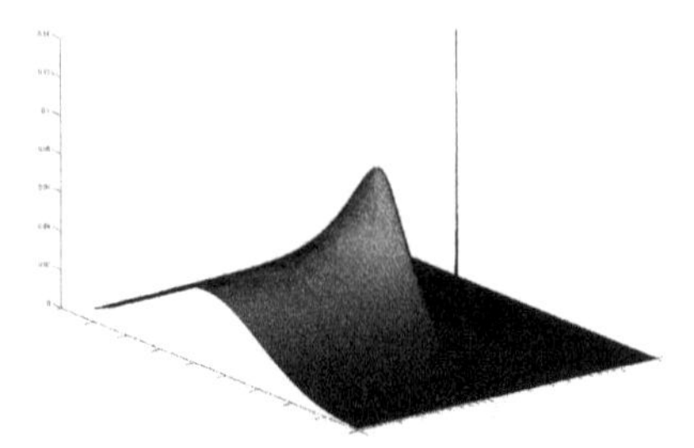

Fig. 7.6 2D cut of a likelihood logarithm

the point G. We cannot say anything about advantages of the second type of a SG manifestation.

Another similar example is an interpolation or an extrapolation of time series. Of course, if we have n points, there is a polynomial of degree $n-1$ interpolating a data set exactly but some polynomial of small degree would be more appropriate solution though it would not be as accurate. For this or other similar examples, the described SG manifestation would be beneficial.

7.4.4 Metaevolutions

Every powerful EC has some parameters, and much experience shows us that there are no "good" parameters for all possible tasks. Certainly, looking for the optimal parameters for a particular task is a task which can be solved by an EC. But metaevolution would be nearly or even totally impracticable if the evolution is time demanding. There is however an opportunity to add EC parameters to the desired solution (as it was done in Sect. 7.4.2), i.e. to make the metaevolution part of the evolution. Above we tried to show that SGs can emerge in an EC if a gene has an opportunity to influence reproduction of its vehicles. Even the selection ruled by a fitness function cannot prevent a SG manifestation. We tried to find benefits of the described SG manifestations. The robustness of solutions was the main benefit that we have found. But in case of metaevolution, robustness of the evolution is not beneficial although robustness of the solution beneficial is (not to mention the frozen evolution). Thus we see the SG role in a metaevolution as negative, i.e., SGs are limiting factors.

7.5 Selfish Genes and Artificial Intelligence

An attempt to express our idea of a relation between SG and embodiment of a logic system is the goal of this section. Construction of neural systems of intelligent animals cannot be fully controlled by their genes because it is obvious that 1) elimination of the influence of the environment seems to be impossible, and 2) there is insufficient amount of information in genes for building such complex structures. One hypothesis which can explain this phenomenon is the hypothesis that these complex structures get around known genetic material and reproduce in some other way. If we ignore this hypothesis the reasoning can continue in the following way: Genes do not rule all building but they leave space for "natural growth". Certainly, this is not a new idea and there is an interesting fact that this idea of "growing" is already described in the play R.U.R. by Karel Čapek published in 1920 in which the term "robot" was coined.[9] We can see the origin of the growth in chaotic behavior of an information system, cf. [20, 21] or Chap. 8 of this book, in some mechanism of self-organization [22], in total lack of causality in living things , or in something completely different.

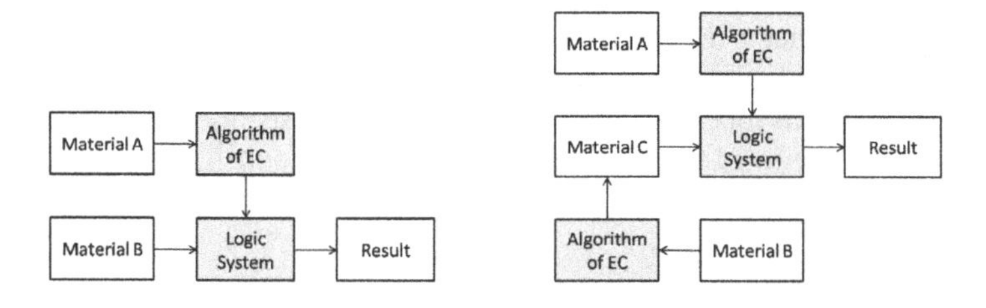

Fig. 7.7 Illustration of the role of ECs in artificial intelligence

Let us imagine that a system of a true artificial intelligence (whatever it means) does not have to (or even cannot) be created by means of an analytical design as a logic system, yet a true artificial intelligence can spontaneously grow up by means a logic system. The term "spontaneously grow" does not mean only emergence on a horizon [23] but it means also creation of an artificial intelligence in the same way as an artificial forest which is not made from artificial trees but it is intentionally made from real natural trees. Hence now, the hypothetical problem would not be construction of an artificial intelligence but construction of a system for an artificial system construction. This situation is showed in the first schema in Figure 7.7. The result constitutes an artificial intelligence system. A similar consideration led us to the metalearning method proposal published in [24]. Naturally, the "material" for the growing cannot be an intelligent system because an artificial intelligence is not supposed to be an analogue of Automaton Chess Player by Wolfgang von Kempelen — instead it must be indifferent and it must have complex behavior with regard to complexity of natural intelligent beings. If the "material" should overcome analytical capacities of a human, it's behaviour must be perfectly unpredictable. A system for growing such intelligence must be very robust, otherwise the growth

[9] "There's something on the inside of them that needs to grow or something. And there are lots of new things on the inside that just aren't there until this time. You see, we need to leave a little space for natural development. And in the meantime the products go through their apprenticeship." [19]

would be an unrepeatable or very improbable event. Only such EC that provides results with both high fitness and robustness can be used for the construction of a system for growing.

Naturally, not every embodiment of a logic system can grow a complex system exceeding analytic abilities. And it is ECs that show how a system of growing might look like. In case of ECs, if we try to differentiate 1) the logic system which operates with a material, and 2) the material where the result grows, we can see 1) an algorithm of an EC, and 2) a random number generator. It is important to realize that the question about manifestation of a SG (i.e. the question of SG presence or absence) is not only a matter of material properties but it is also (or mainly) a matter of the EC algorithm properties. So, SG does not surprisingly originate from the material only. The question "Is this a way or even the only way towards artificial intelligence?" is not answered in this paper because the most important question, i.e. the question "What is the material for growing artificial intelligence?", remains. The material which is used in ECs could be insufficient for this purpose. However, ECs could serve for a suitable material discovery (see the second schema in Fig. 7.7).

7.6 Conclusion

In this paper, we have tried to demonstrate the possibility of an EC design in which the presence of selfish genes can be proved somehow. At first, we have described how this situation can happen. Then we have described how the presence of selfish genes can be proven. Then we have shown some simple examples of ECs with a selfish gene, whose presence we have proved. We have discussed the possible benefits of selfish genes in ECs. Finally, we have described the role of the selfish gene in metaevolution where we see it mostly as a limiting factor.

Acknowledgements. This research was supported by the Ministry of Education of the Czech Republic, project No. MŠMT LC536.

References

1. Vondrejs, V.: What is a gene? Vesmír 2(91), 76–79 (2012)
2. Dawkins, R.: The selfish gene. Popular Science. Oxford University Press (1989)
3. Flegr, J.: Frozen Evolution: Or, That's Not the Way It Is, Mr. Darwin - Farewell to Selfish Gene. Booksurge Llc (2008)
4. Blackmore, S.: The meme machine. Popular Science. Oxford University Press (2000)
5. Tseng, G.C., Wong, W.H.: Tight clustering: A resampling-based approach for identifying stable and tight patterns in data. Biometrics 61(1), 10–16 (2005)
6. Darwin, C.: On the Origin of Species by Means of Natural Selection. Murray (1859)
7. Ridley, M.: Evolution. Blackwell Pub. (2004)
8. Hamilton, W.D.: Extraordinary Sex Ratios. Science 156(3774), 477–488 (1967)
9. Hines, W.: Evolutionary Stable Strategies: A review of basic theory. Theoretical Population Biology 31(2), 195–272 (1987)

10. Altenberg, L.: The evolution of evolvability in genetic programming (1994)
11. Koza, J.R.: Genetic programming (1997)
12. Dawkins, R.: Climbing mount improbable. Penguin science. Penguin (1997)
13. Corno, F., Sonza Reorda, M., Squillero, G.: The selfish gene algorithm: a new evolutionary optimization strategy. In: SAC: ACM Symposium on Applied Computing, pp. 349–355 (1998)
14. Zhang, J., Bushnell, M.L., Agrawal, V.D.: On random pattern generation with the selfish gene algorithm for testing. In: Digital Sequential Circuits, Proc. International Test Conf., pp. 617–626 (2004)
15. Onoyama, T., Maekawa, T., Sakurai, Y., Tsuruta, S., Komoda, N.: Selfish constraint satisfaction genetic algorithm for planning a long-distance transportation network. JCP 3(8), 77–85 (2008)
16. Barta, Z., Flynn, R., Giraldeau, L.A.: Geometry for a selfish foraging group: a genetic algorithm approach. Proceedings of the Royal Society, London, B 264, 1233–1238 (1997)
17. van der Meijs, F.: Implementation of the parallel selfish gene genetic algorithm (2008)
18. Bishop, C.M.: Pattern Recognition and Machine Learning (Information Science and Statistics). Springer-Verlag New York, Inc., Secaucus (2006)
19. Čapek, K., Selver, P., Playfair, N.: R.U.R. Thrift Edition Series. Dover Publications (2001)
20. András, P.: The Role of Brain Chaos. In: Wermter, S., Austin, J., Willshaw, D.J. (eds.) Emergent Neural Computational Architectures Based on Neuroscience. LNCS (LNAI), vol. 2036, pp. 296–310. Springer, Heidelberg (2001)
21. Zelinka, I., Celikovský, S., Richter, H., Chen, G.: Evolutionary Algorithms and Chaotic Systems. SCI. Springer (2010)
22. Kelso, J.: Dynamic patterns: the self-organization of brain and behavior. In: Complex Adaptive Systems. MIT Press (1995)
23. Romportl, J.: Consciousness and causal paradox of emergent systems. In: Interdisciplinary aspects of human-machine co-existence and co-operation, Prague, Czech Technical University, pp. 97–105 (2005)
24. Zelinka, J., Romportl, J., Müller, L.: A Priori and A Posteriori Machine Learning and Nonlinear Artificial Neural Networks. In: Sojka, P., Horák, A., Kopeček, I., Pala, K. (eds.) TSD 2010. LNCS, vol. 6231, pp. 472–479. Springer, Heidelberg (2010)

Chapter 8
Nonlinear Trends in Modern Artificial Intelligence: A New Perspective

Elena N. Benderskaya

Abstract. Artificial intelligent systems capable of learning, setting goals, solving problems, finding new solutions and unforeseen behavior scenarios without external assistance exist in a variety of odd models, each using different sets of assumptions and each having its own limitations. The author argues whether the modern artificial intelligent systems can be called truly intelligent so that to be able to realize the vast capabilities of the human brain. In order to generalize core issues of the main artificial intelligence domains such as fuzzy logics, probabilistic reasoning, bio-inspired techniques, neural networks apparatus, and neuroscience advances together with the new areas of chaos theory and nonlinear dynamics, a multidisciplinary analysis has been employed in the work. In the analysis, the evolution of a particular mathematical apparatus is being considered to justify the application of dynamic models with unstable dynamics used to solve intelligent problems of the next generation. The conclusion is soundly based on the idea that the future of artificial intelligence lies in the sphere of nonlinear dynamics and chaos that is absolutely critical to understanding and modeling cognition processes.

8.1 Introduction

This chapter analyses the evolution of artificial intelligence (AI) paradigms and the transformational role of multidisciplinary research. The leading role of multidisciplinary knowledge is the key to attaining a qualitative leap in AI. Still it is a matter of future investigations to develop the really true human intelligence although a lot of promising results in different AI areas have already been published [14, 27, 38]. A great variety of successful AI applications have found their places in the chapters of this book. This chapter aims to share author's vision on intelligent systems development through accumulated experience in the area. In order to reach this aim we

Elena N. Benderskaya
St. Petersburg State Polytechnical University, Faculty of Computer Science,
Politechnicheskaya 21, 194021 St. Petersburg, Russia
e-mail: helen.bend@gmail.com

J. Kelemen et al. (Eds.): Beyond Artificial Intelligence, TIEI 4, pp. 113–124.
DOI: 10.1007/978-3-642-34422-0_8 © Springer-Verlag Berlin Heidelberg 2013

have to avoid consideration of separate fragmented studies and focus our attention on the creation of a holistic picture of scientific advances in AI field. It is our strong belief that analyzing key issues of widely known theories may give us promising directions in modern AI.

Two main approaches within AI field, namely Symbolic AI and Connectionist AI, no more compete but supplement each other [22, 30]. This statement is confirmed by rapid development of neuro-fuzzy methods combining both learning processes and explicit knowledge statements.

However, despite this mutual enrichment AI theory is still far from its ambitious goal that seemed to be so quick and easy to reach. Genuine intelligent systems understood as capable to learn, set goals, solve problems, find new solutions and unforeseen behavior scenarios without external assistance exist in the form of separate models functioning under a wide range of assumptions and limitations [3, 6, 7, 31]. The question left is whether existing artificial intelligent systems can be called truly intelligent systems fully describing the multiformity of Universe.

The discussion of the nature and essence of Intelligence is becoming more and more holistic. Some aspects of this complex notion are being discussed in the book. Although philosophical uncertainty and social responsibility issues are often left behind the actual AI research, the importance of the particular issues is of no question at all.

8.2 Modern Artificial Intelligence

Since the moment of the origin, artificial intelligence has been a multidisciplinary field and comprised knowledge from various scientific domains [27]. It is likely to become a starting point for mutual penetration of sciences, the process that is quite opposite to sciences differentiation.

Though the need for interdisciplinary investigations sufficiently increased, the opposite processes of sciences differentiation are flourishing due to the complexity of the research objects and the intricacy of theories.

8.2.1 Limitations of Basic AI Approaches

There is a huge amount of isolated single-purpose AI models and methods [27, 38] effective to solve particular narrowly defined problems. However, it is hard to consider them as basic ideas for a general theory. There are a lot of theoretical contributions and empirical material on the subject of artificial intelligence, but the overall picture is still very fuzzy.

It is obvious that the idea of combining all existing artificial intelligence models into a super one won't yield any benefits. Thus there arises the choice issue of most perspective trend in artificial intelligence theory.

One of the main issues challenging for Symbolic AI researchers includes the impossibility of formalizing the representation of all situations which a system may encounter during functioning [30, 31]. After making some assumptions, a set of certain problems (dependent on the limitations) is successfully being solved. Nevertheless, universality still appears unattainable. The system is intelligent to the extent predetermined by input data, comprehensiveness, and predesigned scenarios. Be it as it may, intelligence implies the generation of new knowledge inside the system even if the input data is incomplete or contradictory

Connectionist AI researchers face restrictions representing the adequate size and quality of training samples. For example, neural networks (NN) are good at operating with implicit data and generalizing information through learning process. However, approximation results of feed-forward networks rely [7, 20] on the quality of training data. In the real world, comprehensive data in most cases is unavailable.

Briefly speaking, classical NN represent the parallel implementation of corresponding pattern recognition methods based on algebraic or probabilistic approaches. Thus these methods inherit advantages and most of disadvantages of the approaches.

8.2.2 Evolution of Formal Methods: Dealing with Uncertainty

Symbolic AI responds to uncertainty issues in four different ways reflected by the enrichment of mathematical apparatus. We know that classical and widely applied mathematics is based on the usage of either discrete or continuous variables. Digital devices use discrete data representation, while analogous equipment operates with continuous values.

In both cases separate numbers, elements, and points are used. Classical mathematics is dominated by point attractors. Classical computing architectures work with accurately defined input data according to predetermined algorithms. In hardly formalized problems, this requirement is not so easy to fulfill, as real data is only partially truthful. It is also incomplete, redundant, and imprecise.

A substantial breakthrough in managing linguistic uncertainty is considered to be the shift to fuzzy sets and fuzzy logic development. It produces opportunities to process intervals and ranges of values. As the main operation component, the point-number value is replaced by the interval one.

Nowadays, further development of fuzzy sets and fuzzy logic is being undertaken in an extensive way. To operate intervals within intervals, that is to operate within an additional fuzzy dimension, fuzzy sets type 2 has been designed. On the one hand, this apparatus helps to describe intricate input data sets, on the other hand, the apparatus of fuzzy sets type 2 gives us less research significance in comparison to the shift from point logic to interval type 1 logic. In case of type 2 fuzzy logic, one faces a substantial increase in computational complexity which stands to become incommensurable compared to modest growth of solutions effectiveness.

Fuzzy computers and neuro-fuzzy hybrid solutions were developed to accomplish specific fuzzy logic operations. Hybrid systems advantageously combine fuzzy data representation with the learning capacities of neural networks [22].

It is worth noting here that the fuzzy logic apparatus is related to knowledge bases and the expert approach of AI domain. From this perspective, fuzzy logic is incorporated into inference engines and production rules. The expert opinion specifies membership functions for all considered input data combinations. To develop an expert system, a knowledge engineer is required to formalize the problem in the form of production rules combinations. In this case, the subjective opinion of an expert, the core of an expert system, is one of the principal weaknesses.

We suppose that probabilistic reasoning may be considered the way with the help of which uncertainty can be overcomed. Managing statistical type of uncertainty, probabilistic reasoning uses separate numerals the nature of which is probabilistic.

The result of probabilistic reasoning depends greatly on the a-priori information about the possible variety of distributions. Thus a lot of hypothesis should be tested. However, in this case there is still no guarantee that the final estimations will be closer to real ones. The application of probabilistic reasoning is seriously constrained by the level of problem examination.

The next step in research efforts to enrich computations throughout the real-world multivariate answers is made using chaos theory [39]. Chaotic systems predetermine the emergence of structures containing all the required diversity in answers. In comparison to quantum computations, nonlinear systems are somewhat easier to deal with, and thus more appropriate for observation and analysis.

The behavioral regimes produced by chaotic systems are split into positive and negative aspects. The positive ones deal with optimistic assumptions about rich systems dynamics encompassing problems solutions of any kind. The negative ones speak about the control problems in such systems. Together with desired useful behavior, chaotic systems are very vulnerable and can be observed in absolutely opposite regimes.

To define useful chaos, one must first define the concept of "deterministic chaos". The behavior of deterministic chaotic systems is reproduced from experiment to experiment. Thus it is possible to analyze and apply chaotic systems for solving problems. Deterministic chaos is identified on the border of order and turbulence. Order predetermines the existence of structure and thus the convergence to solution, in other words some stable behavioral regime. These regimes induced by the emergence of self-organized structures take place only when a lot of nonlinear elements coexist in systems dynamics. Chaotic behavior is generated by both cooperative and individual dynamics of elements. This fact correlates greatly with recent advances in hardware implementation of complex programmable logic architectures discussed later.

Mathematics of nonlinear dynamics and chaos are considered to be the next stage in the development of a mathematical apparatus. At the previous stages one starts with deterministic models and goes to stochastic models, then from stochastic models to deterministic chaotic models. The evolution of mathematical models is given on figure schematically (Fig. 8.1). It depends on the research systems types (static or dynamic) and the effects that could take place (type of attractors they converge to). The scheme does not contain information about the time of models emergence. This is done on purpose in order to underline the influence of modern computer modeling opportunities unavailable sometime ago, when mathematical models were first introduced. As systems with chaotic dynamics possessing chances to describe

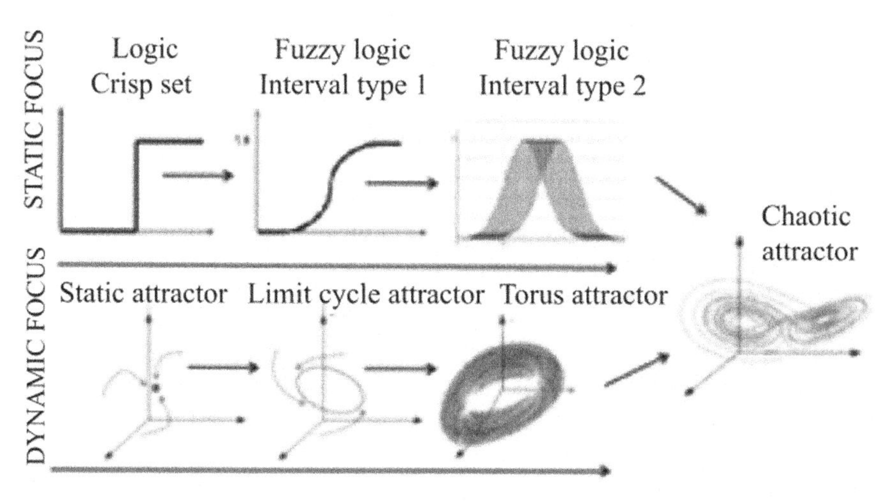

Fig. 8.1 Evolution of formal methods: dealing with uncertainty

multiformity of Universe, it is important to consider this perspective in relation to the recent advances in neuroscience.

8.3 Neuroscience: Brain as Chaotic Computer

Due to incomparable abilities of intelligent systems in decision making, it is natural to investigate brain activity, its texture, and functioning principles. Thanks to modern techniques in non-invasive MRI of brain cortex, terabytes with brain functioning evidences are now accumulated. Recent research in the field of genetic engineering and broad experimental results on brain dynamics has brought much valuable knowledge on the activities of the cortex [23, 25, 41]. High performance computing makes it possible to model and examine in detail the memory and behavioral-related processes brain cells go through.

A lot of investigations deal with the estimation of chemical compound concentrations and the changes in electrical activities that indicate different states of neural systems [14, 23, 25]. It is worth mentioning that neural equations describing neuron functioning were introduced in the late 1940s. At that time, the researchers accepted the idea that having knowledge of the structure and properties of basic construction elements gave them the possibility of extending this knowledge in order to create similar artificial intelligent system. Not much progress has been made in the meantime. Only recently people have obtained results modeling a part of rat brain functioning. Biological neurons and networks on substrates (biological chips) have been also cultivated [2, 26]. It's time to integrate all practical AI knowledges and the knowledges of relative scientific fields within the framework of analysis and synergetics.

Brain activity is measured by different techniques. Most of them including electroencephalogram technology give us evidence about the presence of chaos in brain dynamics [11, 12]. More over, chaotic oscillations based on neuroscience results characterize healthy brain dynamics. To find out the role of chaotic dynamics in pattern recognition processes a researcher goes through evolutionary stages: from neural network to a separate neuron, from a separate neuron to a DNA molecule. Although these results explain the mechanism of memory consolidation, still there is a long way to understanding of upper level cognition processes [19].

Wide knowledge about systems functioning at the micro-level seldom leads to the perception of a system as a whole entity. Systemic effects are generated not in a small part of a system. It is generated by the sum of those system components. It is the holistic thinking that helps to see and understand a big picture of a system and the process of interaction within it [7, 17, 31, 36].

Therefore, the dilemma can be stated as a compromise between the extent of biological processes imitation and the bearable level of abstractions from the natural prototype.

All the ideas discussed above make us focus our attention on an integrator of trans-disciplinary knowledge of complex systems and model brain dynamics. From the author's point of view, the best illustration of the tool is offered by the self-organization concept. It is a well known fact that the phenomenon of self-organization is the core issue examined by synergetics, the discipline that is highly correlated with nonlinear dynamics and chaos theory [17, 36]. The necessity of paying attention to self-organization patterns and models is mentioned in Chapter "Multi-agent Systems in Industry: Current Trends and Future Challenges" of this book. It seems to be quite reasonable the application of synergetics to the field of human brain functioning, as human brain itself is recognized to be a complex nonlinear system comprising millions of globally coupled neurons interacting by means of electro-chemical signals. An important implication of the research efforts undertaken in this direction is the substantial growth of interest to bio-inspired and dynamic approaches in different scientific domains.

8.4 Novel Bio-inspired Methods and Dynamic Neural Networks

In spite of the fact that the idea of artificial intelligence is itself bio-inspired, the reproduction of biological prototype is accomplished differently in the process of AI development [14]. Classical NN models consist of formal neurons. Recent investigations are paying more and more attention to bio-inspired NN with neurons similar to nerve cells. They touch upon the exhaustive phase in NN models design, because the quality improvement is much less than amount of efforts spent on the research in the area. Nevertheless, without better understanding of the principles of the mutual interactions, a detailed reproduction of the processes occurring in separate nerve cells will likely yield disappointments.

The results of applying bio-inspired methods only supplement our knowledge of multiform behavior of biologic systems [13] whereas the simple reproduction of neuron ensembles characteristics for partial regimes is unlikely to be generalized. The improvement of one set of parameters negatively influences the other one previously well-tuned. It is doubtful that even bottom-up design starting from the lowest level may result in the emergence of a system with new qualities since the synergy effects occur mainly when self-organization principals underlie the system [17, 42].

It is a well known fact that there are two classes of NN. Nowadays, nearly all novel results are obtained in the area of static NN. At the same time, the interest towards dynamic NN is constantly growing up. To prove it, the evolution of dynamic NN can be presented in the following stages:

1. The first stage: the appearance of Hopfields and Hakens neural networks attractor neural networks. Attractors in the first dynamic neural network are simple enough fix-point attractors [16, 20].
2. The second stage: dynamic neural networks with more complex cycle attractors and torus attractors [17, 45].
3. The third stage (current): dynamic neural networks with unstable dynamics most complex chaotic attractors. Their dynamic is characterized by a set of trajectories scaled in the phase space to a location with an infinite number of switching states [4, 29, 44].

The multivariate nature of desired systems behavior is the starting point in designing the basis for order/attention parameters formation introduced by Haken in [16]. Thus, the most important context issues and situation preferences can be taken into account (see Chapter "Analogy, Aesthetics, and Affect: What HCI Designers Can Learn from AI" in this book).

A separate sub-class of modern dynamic neural networks is a set of models of reservoir computing [28, 29]. The origins of such systems can be found at the second stage as they have random structure, and they are also capable to generate promising complex dynamics [1].

The idea of implementing a stochastic component so that to extend freedom dimensions was proposed long ago and more than once [15]. Together with the investigation of the amazing capabilities of nonlinear systems, assumptions on a mathematical apparatus adequately describing the complex behavior of large-dimension dynamic systems were also made. All over the world, the research spiral is returning now and again to chaos applications. However, the lack of an appropriate hardware basis has become an obstacle to further development of chaos implementation.

Recurrent neural networks include neurons of different complexity. Building blocks can be very simple; they may be like threshold units or some other formal models. However, the simple blocks can imitate the real cellular activity. In this case, Hodgkin-Huxley neurons can be used. The dynamics they produce can be of very different types: either regular or spiking. Therefore, at the level of the whole reservoir, not at the micro level of a neuron, the dynamics is very complex.

An important issue raised by researchers while discussing the problem of reservoir computing is the issue of unguaranteed desired dynamics since there is much uncertainty about the reservoir parameters, speaking about Maass bio-inspired model in particular. Assignment of parameters is accomplished in empirical and intuitive manner on the basis of preliminary experiments. There is a body of evidence to suggest that a Turing machine could be developed on the base of reservoir neural networks [29], but a practical application of this approach is still a matter of future investigations.

The existence of a great number of parameters and variables in complex systems leads to the variety of regimes. The phenomenon can be interpreted in the terms of freedom degrees. It is well known that rich dynamics is observed in the systems with a large number of freedom degrees. Rich dynamics defines different types of systems behavior namely ordered, partially ordered, and turbulent. For the purpose of solving artificial intelligent problems, partially ordered regimes are considered to be more adapted.

Unfortunately, it is not so easy to find an appropriate combination of parameters to obtain partially ordered or ordered regimes. Besides, there are also disadvantages in plasticity. It is quite obvious that a lot of difficulties in control appear in this case; and only a specific combination of parameters may give us synchronized ordered regimes. But the problem is that there is no information about how to find these intervals.

8.5 Chaotic Bio-inspired Systems

Nowadays, a lot of research in the field of micro level description of neurons, synapses, molecules, and atoms has been conducted. Still, there is a wide gap between the activation of a neuron and the description of real cognition processes in the brain. There is a kind of "Grand canyon" between the micro level of representation resembling one drop of water and the macro level resembling a tsunami catastrophe. As it is a hard job to make a leap from the micro to macro level, new paths for analysis and synthesis should be considered [19].

We may need something simpler than structural and dynamical complexity of units and systems. We suppose that all the difficulties aim at producing order from chaos by means of synchronization and resonance effects in bio-inspired systems.

The most promising paths on the way existing in AI area are dynamic self-organization, synchronization, resonance effects, nonlinear (chaotic) dynamics, and chaotic bio-inspired systems [18, 24, 37, 40]. Another approach is to produce complex dynamics by a simple way.

When developing a network with some complex dynamics, basic transfer functions in recurrent NN are replaced with chaotic maps [4, 32]. Growth of complexity goes in the directions of increasing intricacy in processing unit linkage and complication of units themselves. Chaotic maps generate deterministic chaos and thus their application combines both trends. One of the most applicable map is a logistic map that allows controlling chaos by means of one parameter.

Chaotic systems dynamics depends strongly on external circumstances and thus can help to represent the whole context as well as a variety of possible situations understanding as a mobile and preliminary undefined environment in which tasks might be solved. Some promising results on the issue have already been presented in a number of publications [5, 8, 17, 29, 35, 36, 42, 45].

Nevertheless, the development of a unified approach to AI requires the examination of the system functioning in the context of the environment. This idea is being widely discussed [35] with the focus on the role of chaos agent development. In the long run, it is the holistic approach that will likely help to formalize this complex notion of context. The significance of context in problem solving and knowledge structuring is being discussed in Chapter "Beyond Knowledge Systems" of this book.

In order to link an intelligent system with the environment, it is necessary to apply high-quality sensors. This is an essential condition of not only the adequate perception of the real world but also of the proper design of associations and historical knowledge (Chapter "Beyond Knowledge Systems" of this book).

8.6 Chaotic Computing

When powerful computer equipment had been designed, it became possible to model chaotic systems. Numerous experiments have been made, and at present rather precise solutions speaking for the relation between chaos effects and non-linear transfers in systems can be obtained.

One of the best illustrations of chaos application is the synergy of informatics and physics. For several decades physicists has analyzed the dynamics of coupled logistic maps generating cooperative behavior in the form of synchronized clusters. One of the most challenging tasks in data mining is clustering without a-priory information about the structure of patterns in the data. Chaotic neural network gives an opportunity to develop clustering systems of high quality [5]. It becomes possible thanks to a combination of metric heuristics and new computational structures based on chaotic transfer units.

Besides chaos application to clustering problems, there have also appeared successes in the information transfer. The main aim of chaos application in this area can be identified as data encryption within information security systems [9, 43]; steganography (information concealment), wideband signals generation [10]; and weak radio-signals detection in radio-location systems [33, 34]. Thus, a new scientific domain has come into being, so called chaotic computing, chaotic processors and chaotic computers being currently under focus of broad research.

The dissemination of chaotic systems is restrained, on the one hand, by complicated approaches to nonlinear systems analysis and, on the other hand, by the necessity of applying not only inter-disciplinary knowledge but also trans-disciplinary awareness for the purposes of developing chaotic intellectual systems.

Although there exist digital and analogous devices with chaos namely chaotic processors and chaos generators [8], at the moment there is a significant demand for a cheap and unified hardware support of chaotic computing.

The most appropriate hardware platform for chaotic computing seems to be analogous systems. The idea that the future lies in analogous computations and thus effective neural networks computers should be realized on analogous basis is being discussed in publications [21]. Analogous computers operate with continuous signals which contain an infinite variety of input data and solution combinations. This particular fact becomes crucial for the consideration of chaotic systems since by nature they have a wide range of behavioral regimes.

The analysis of hardware development trends allows concluding that the neural networks paradigm is the apparatus mostly for implementing intellectual functions. Though the power of modern computers increases rapidly, there is a demand for high performance computing to solve real-life intelligent problems. Therefore, it seems reasonable to apply naturally parallelized neural networks.

8.7 Conclusions

The chapter considers core issues of main AI domains such as fuzzy logics, probabilistic reasoning, bio-inspired methods, and neuroscience advances together with new adjoining areas of chaos theory and nonlinear dynamics. Any attempt in investigating some cortex domain via detailed reproduction of neuron cell ensembles is a rational one in neurophysiology's micro-level models. However, there can be doubts as to the merit of this approach when artificial intelligence in the macro-level perspective is considered. A unified approach to the development of AI systems with the quality comparable with natural neural systems is proposed in order to generate the distributed ensembles of chaotic coupled maps. This direction of research is attractive for it combines the ideas of an agent theory, neural network apparatus, nonlinear dynamics, synchronization theory, informatics, and formal logics. Thus, hybrid solutions of new generation come into being.

The above-stated discussion leads us to conclude that it is high time we should unite knowledges gained from multidisciplinary research projects in order to solve commensurably complicated intelligent problems.

Consequently, a wide range of opportunities to apply complex synergistic effects to the problem of uncertainty not only in technical but also in biological, economic, geopolitical systems strengthens the view that very soon nonlinear dynamics and chaos will become the most demanded apparatus to understand and model cognition processes.

Conceived in the last century, the nonlinear approach now opens up new perspectives for intelligent systems development. These opportunities are predetermined by a decrease of known limitations and growth of solutions universality. Also, this is a result of the multidisciplinary approach and sciences integration. Synergetics as a modern philosophy of sciences puts system analysis on the principal position in the experience of benefits of systems integration and self-organization.

References

1. Andreyev, Y.V., Dmitriev, A.S., Chua, L.O., Wu, C.W.: Associative and random access memory using one-dimensional maps. International Journal of Bifurcation and Chaos 2(3), 483–504 (1992)

2. Baruchi, I., Ben-Jacob, E.: Towards Neuro-Memory Chip: Imprinting Multiple Memories in Cultured Neural Networks. Physical Review E 75, 50901 (2007)
3. Baum, S.D., Goertzel, B., Goertzel, T.: How long until human-level AI? Results from an expert assessment. Technological Forecasting and Social Change 78, 185–195 (2011)
4. Benderskaya, E.N., Zhukova, S.V.: Fragmentary Synchronization in Chaotic Neural Network and Data Mining. In: Corchado, E., Wu, X., Oja, E., Herrero, Á., Baruque, B. (eds.) HAIS 2009. LNCS (LNAI), vol. 5572, pp. 319–326. Springer, Heidelberg (2009)
5. Benderskaya, E.N., Zhukova, S.V.: Oscillatory Chaotic Neural Network as a Hybrid System for Pattern Recognition. In: IEEE SSCI 2011 - Symposium Series on Computational Intelligence - HIMA 2011: 2011 IEEE Workshop on Hybrid Intelligent Models and Applications, pp. 39–45 (2011)
6. Bobrow, D.G., Brady, M.: Artificial Intelligence 40 years later. Artificial Intelligence 103, 1–4 (1998)
7. Cristianini, N.: Are we still there? Neural Networks 23, 466–470 (2009)
8. Ditto, W.L., Murali, K., Sinha, S.: Chaos computing: ideas and implementations. Phil. Trans. R. Soc. A 366, 653–664 (2008)
9. Dmitriev, A.S., Kyarginsky, B.Y., Panas, A.I., Starkov, S.O.: Direct chaotic communications schemes in microwave band. J. Commun. Technol. Electron. 46, 224–233 (2001)
10. Dmitriev, A.S., Efremova, E.V., Kletsov, A.V., Kuzmin, L.V., Laktyushkin, A.M., Yu, Y.V.: Ultra Wideband Signals and Sensor Networks. In: Proc. 1st Int. Conf. Management of Technologies and Information Security (ICMIS 2010), Allahabad, India, January 21-24, pp. 48–64 (2010)
11. Freeman, W.J., Skarda, C.A.: Spatial EEG patterns, nonlinear dynamics and perception: the neo-Sherringtonian view. Brain Res. 357, 147–175 (1985)
12. Freeman, W.J.: Neurodynamics: an exploration in mesoscopic brain dynamics. In: Perspectives in Neural Computing. Springer (2000)
13. Freeman, W.J.: Evidence from human scalp electroencephalograms of global chaotic itinerancy. Chaos 13(3), 1067–1077 (2003)
14. Goertzel, B., Ruiting, L., Itamar, A., Hugo, G., Shuo, C.: A world survey of artificial brain projects, PartII: Biologically inspired cognitive architectures. Neurocomputing 74, 30–49 (2010)
15. Granichin, O., Gurevich, L., Vakhitov, A.: Discrete-time minimum tracking based on stochastic approximation algorithm with randomized differences. In: Proceedings of the Combined 48th IEEE Conference on Decision and Control and 28th Chinese Control Conference, Shanghai, pp. 5763–5767 (2009)
16. Haken, H.: Neural and Synergetic Computers. Springer (1988)
17. Haken, H.: Synergetic Computers and Cognition: A Top-Down Approach to Neural Nets. SSS. Springer (2010)
18. Havel, I.M.: Artificial Intelligence and Connectionism: Some Philosophical Implications. LNCS (LNAI), pp. 25–41. Springer, Heidelberg (1992)
19. Havel, I.M.: Causal Domains and Emergent Rationality. In: Proceedings of the 23rd International Wittgenstein Symposium, Vienna, pp. 129–151 (2001)
20. Haykin, S.: Neural Networks: A Comprehensive Foundation. Prentice-Hall (1998)
21. Horio, Y., Aihara, K.: Analog computation through high-dimensional physical chaotic neuro-dynamics. Physica D: Nonlinear Phenomena 237(9), 1215–1225 (2008)
22. Jang, J.R., Sun, C., Mizutani, E.: Neuro-Fuzzy and Soft Computing: A Computational Approach to Learning and Machine Intelligence. Prentice Hall (1997)
23. Kazanovich, Y.B.: Nonlinear dynamics modeling and information processing in the brain. Optical Memory and Neural Networks 16(3), 111–124 (2007)

24. Kay, L.M.: Olfactory Coding: Random Scents Make Sense. Current Biology 21(22), R928–R929 (2011)
25. Korn, H., Faure, P.: Is there chaos in the brain? I. Concepts of nonlinear dynamics and methods of investigation. Life Sciences 324, 773–793 (2001)
26. Li, N., Tourovskaia, A., Folch, A.: Biology on a chip: microfabrication for studying the behavior of cultured cells. Critical Reviews in Biomedical Engineering 31, 423–488 (2003)
27. Luger, G.F.: Artificial Intelligence: Structures and Strategies for Complex Problem Solving. Addison Wesley (2008)
28. Lukoviius, M., Jaeger, H.: Reservoir computing approaches to recurrent neural network training. Computer Science Review 3(3), 127–149 (2009)
29. Maass, W., Natschlaeger, T., Markram, H.: Real-time computing without stable states: A new framework for neural computation based on perturbations. Neural Computation 14(11), 2531–2560 (2002)
30. Mira, J.M.: Symbols versus connections: 50 years of artificial intelligence. Neurocomputing 71, 671–680 (2008)
31. Oliveira, F.: Limitations of learning in automata-based systems. European Journal of Operational Research 203, 684–691 (2009)
32. Pikovsky, A., Rosenblum, M., Kurths, J.: Synchronization: A Universal Concept in Nonlinear Sciences. Cambridge University Press, CNSS (2003)
33. Potapov, A.A., German, V.A.: Detection of Artificial Objects with Fractal Signatures. Pattern Recognition and Image Analysis 8(2), 226–229 (1998)
34. Potapov, A.A.: The Textures, Fractal, Scaling Effects and Fractional Operators as a Basis of New Methods of Information Processing and Fractal Radio Systems Designing. In: Proc. SPIE, vol. 7374, pp. 73740E-1–73740E-14 (2009)
35. Potapov, A.V., Ali, M.K.: Nonlinear dynamics and chaos in information processing neural networks. Differential Equations and Dynamical Systems 9(3-4), 259–319 (2009)
36. Prigogine, I.: The End of Certainty. Free Press (1997)
37. Rojas-Lbano, D., Kay, L.M.: Olfactory system gamma oscillations: the physiological dissection of a cognitive neural system. Cognitive Neurodynamics 2(3), 179–194 (2008)
38. Russell, S., Norvig, P.: Artificial Intelligence: A Modern Approach. Prentice Hall (2002)
39. Sinha, S., Ditto, W.L.: Computing with distributed chaos. Phys. Rev. E 59, 365–377 (1999)
40. Varela, F.: Resonant cell assemblies: A new approach to cognitive functioning and neuronal synchrony. Biological Research 28, 81–95 (1995)
41. Velazquez, J.: Brain, behaviour and mathematics: Are we using the right approaches? Physica D 212, 161–182 (2005)
42. Wolfram, S.A.: A New Kind of Science. Wolfram Media (2002)
43. Yang, T.: A survey of chaotic secure communication systems. Int. J. Comput. Cognit. 2, 81–130 (2004)
44. Zak, M.: An unpredictable-dynamics approach to neural intelligence. IEEE Expert: Intelligent Systems and Their Applications archive 6(4), 4–10 (1991)
45. Zak, M.: Quantum-inspired resonance for associative memory. Chaos, Solitons and Fractals 41, 2306–2312 (2009)

Chapter 9
Membrane Computing in Robotics

Ana Brânduşa Pavel, Cristian Ioan Vasile, and Ioan Dumitrache

Abstract. This paper presents a new computational paradigm which can be successfully applied in robotics for the control of autonomous mobile robots. Membrane computing is a naturally parallel and distributed model of computation inspired by the structure and functioning of living cells. Numerical P systems, a type of membrane systems which operates with numerical values, and the extension, enzymatic numerical P systems, were used for modeling robot behaviors. Current results and developments of this innovative approach are also discussed and analyzed.

9.1 Membrane Computing—An Overview

Membrane computing is an interdisciplinary area that combines mathematics, computer science and biology. It focuses on computational models inspired by the properties and processing mechanisms of biological cells and tissues. The biological cell is an amazing machinery which is capable of processing great amounts of nutrients and bio-molecules in parallel. Membrane systems as a computing device were introduced by the mathematician Gheorghe Paun and were named "P Systems" after his name.

Although P systems were first investigated as a computational model which offered the possibility of solving NP-complete problems in less than exponential time, other perspectives of using this model were proposed and studied. For instance, P systems were applied in biology to model genetic networks for gene expression analysis [10] and metabolic processes [11], [12]. They were also used to model ecological systems and to predict the evolution of different species [3], [4]. We present and discuss a new branch of applications for P systems which can be also used to

Ana Brânduşa Pavel · Cristian Ioan Vasile · Ioan Dumitrache
Department of Automatic Control and Systems Engineering,
Politehnica University of Bucharest, Splaiul Intependenţei, No. 313, 060042,
Bucharest, Romania
e-mail: {apavel, cvasile, idumitrache}@ics.pub.ro

J. Kelemen et al. (Eds.): Beyond Artificial Intelligence, TIEI 4, pp. 125–135.
DOI: 10.1007/978-3-642-34422-0_9 © Springer-Verlag Berlin Heidelberg 2013

model and implement behaviors for autonomous robots. We consider P systems a computational paradigm which can be successfully applied in the fascinating world of cognitive robotics.

A P systems is a compartmental model which can also be represented as a tree-like structure. As a great advantage of the model we mention its parallel and distributed nature. Therefore, computations are made in parallel in each compartment. The components of a P systems can be identified in figure 9.1. The system has a skin membrane placed in an environment. The skin membrane contains inner membranes. If a inner membrane doesn't contain other membranes, it is called an elementary membrane. Each membrane delimits its inner space named region. Each membrane has a label represented here by letter M and a corresponding number.

Currently there exist many types and classifications of P systems [18]. We will consider here the classification in two major types of P systems which are symbolical P systems (SPS) and numerical P systems (NPS). Although most of the research effort in membrane computing has been focused on SPS so far, the advantages of using NPS for solving problems in robotics will be further discussed.

As a general definition, SPS is an universal computational model, equivalent to a Turing machine. The compartments of the membrane system contain multi-sets of symbols and rewriting rules that transform and transport the symbols within the compartments. The symbols can be created or destroyed by the rules and the membranes can be dissolved during the computational process. The mechanisms which lie beyond the transformation of the symbols are inspired by the biochemical reactions which take place inside living cells. The computation of a SPS takes place until no change can be performed within the system. Details about SPS can be found in [17], [18].

NPS are a type of P systems, inspired by the cell structure, in which numerical variables evolve inside the compartments by means of programs; a program (or rule) is composed of a production function and a repartition protocol. The variables have a given initial value and the production function is a multivariate polynomial. The value of the production function for the current values of the variables is distributed among variables in certain compartments according to a repartition protocol. Formal definition of NPS can be found in [19] where the authors introduce this type of P systems with possible applications in economics.

In figure 9.1 a simple NPS example is illustrated in order to show how these systems work. Each membrane contains a number of variables which can store real numbers and a set of rules. In this example, each membrane contains only one rule. The production functions are multivariate polynomials F, G, H. The initial values of the variables are represented between brackets. The Greek letters (α, β, γ, δ, ε, η, θ, λ, μ – see figure 9.1) are constants and represent the coefficients of the repartition protocols. For instance, variable x, which belongs to membrane $M1$ will be updated in one computational step as follows:

$$x \leftarrow x + \alpha \cdot \frac{F(x,y,z)}{\alpha+\beta+\gamma} + \delta \cdot \frac{G(u,v)}{\delta+\varepsilon+\eta}. \tag{9.1}$$

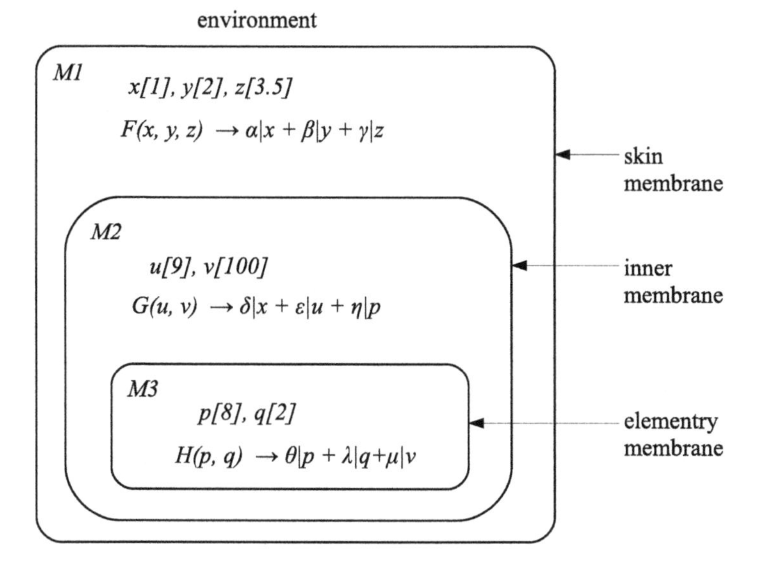

Fig. 9.1 NPS example

A production function which belongs to membrane i may depend only of some variables which belong to membrane i. The variables which appear in the production function become 0 after the execution of the program. If a variable belongs to membrane i, it can appear in the repartition protocol of the parent membrane of i and also in the repartition protocol of the children membranes of i. After applying all the rules in parallel, if a variable receives such contributions from several neighboring compartments, then they are added in order to produce the next value of the variable.

Numerical P systems were designed both as deterministic and non-deterministic systems [19]. Non-deterministic NPS allow the existence of more rules per each membrane and the best rule is selected by an "oracle", while the deterministic NPS can have only one or no rule per each membrane. Designing robot behaviors requires deterministic mechanisms. Therefore, an extension of NPS, enzymatic numerical P systems (ENPS), in which enzyme-like variables allow the existence of more than one program (rule) in each membrane while keeping the deterministic nature of the system were introduced in [15]. Therefore, ENPS are a more powerful modeling tool for robot behaviors than classical NPS. In section 9.3 a comparison between robotic membrane controllers designed with classical NPS [2] and designed with ENPS [14], [16] is presented.

9.2 Membrane Controllers for Mobile Robots

The main advantages of using NPS as a modeling tool are the numerical and the naturally parallel and distributed nature of the model. Membranes of a NPS can be distributed over a grid or over a network of microcontrollers in a robot. The computation done in each membrane region (the execution of the membranes programs) can also be done in parallel. This is very important, because a membrane system, which is an abstract implementation of a desired behavior, can be executed in a distributed and parallel way without having to worry about the design and implementation issues regarding parallelization and distribution. Therefore, NPS can be used as a modeling tool for parallel and distributed control systems.

There are two main advantages of using NPS instead of SPS for modeling robot controllers. Firstly, the variables can receive real numbers, therefore no special effort is needed in order to implement operations on numbers (like floating point operation, for example). Secondly, the set of variables is defined at the beginning of the program and no other symbols may occur during the computation as it happens in SPS; only the values of the variables evolve, but the memory used for storing the variables has a fixed size;

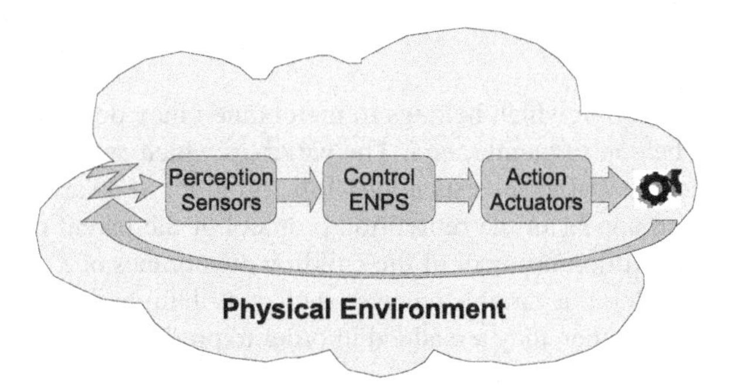

Fig. 9.2 General model of a physical agent

Figure 9.2 illustrates the general model of a physical agent which interacts with the environment. This model is also known as the Observe-Decide-Act loop (ODA) which is also used to represent self-adaptive computing systems [8]. The design of the algorithms used for the control of robots is fundamentally different from the design of the algorithms that solve problems in which all data is well defined and completely known. The world model in which the robot has to act only approximates the physical laws. The world is only partially known and imperfectly modeled. Therefore, the robot controllers have to be designed to overcome uncertainty, errors in modeling and measurement, noise. The authors discuss here the possibility of using NPS as a new computational paradigm for implementing such robot controllers.

In [2], the authors present examples of robot behaviors implemented with NPS for obstacle avoidance, wall following, following another robot. The controller's performance is measured by the mean execution time of a cycle. The cycle represent the computation of a loop in which the robotic system reads the information from the sensors, computes the speeds of the motors based on the information received from the sensors, and sets the new values of the motors' speeds (figure 9.2).

After designing a membrane controller, the membrane system can be simulated using a numerical P systems simulator such as SNUPS [1], SimP [13], or the parallelized GPU-based simulator for ENPS proposed in [7]. One benefit of using membrane controllers is that the computational performance can be increased by improving the simulator's performance (parallelization, distribution and other optimizations) and not by modifying the membrane controllers themselves. The membrane simulator can be considered as a virtual machine like Java or Python virtual machines. It can be seen as a "membrane computer" that runs "membrane programs" (in this case, the membrane controllers). The simulator is the middleware between the hardware and the membrane controllers. Therefore, by optimizing its performance, the performance of all the defined membrane controllers increases.

Design and implementation of robot controllers require deterministic computational models. In order to be deterministic, a NPS model must have only one rule per membrane and, most of the time, this restriction makes the model rigid and difficult to use. Therefore, an extension of the NPS model, Enzymatic Numerical P Systems (ENPS), was proposed in the context of modeling robot behaviors. ENPS model allows the parallel execution of more rules (programs) per membrane while keeping the deterministic behavior. ENPS use some special variables, inspired by the behavior of biological enzymes which, associated to rules (in analogy to chemical reactions), can decide whether a rule is active or not at a given computational step. A rule is active if the associated enzyme has a greater value than the minimum of the variables involved in the rule or if the rule has no associated enzyme. Details about ENPS model together with formal definition and examples can be found in [14], [16] and [15].

Theoretical results about universality of the NPS model are presented in [19] and [21]. In [21], the authors prove that ENPS is an universal model, improving the universality results of the classical NPS model. Therefore, ENPS model can be successfully used as a computing device for modeling and simulating physical processes.

By adding enzyme-like variables to the NPS model, the modeling power of NPS increases. The enzymatic mechanism controls the execution flow of a NPS with multiple rules per membrane. The possibility of selecting and executing more production functions per membrane makes ENPS a more flexible modeling tool than classical NPS. ENPS robot controllers (as those described in [14], [16]) have a less complex structure than the NPS ones (described in [1]), therefore less computations are performed and the performance of the systems increases.

9.3 An Overview of the Results

Both NPS and ENPS models could be used for modeling autonomous mobile robot behaviors [2], [14] and [16]. The numerical nature, the distributed and parallel structure and the computing power, make membrane controllers suitable candidates for the control of complex systems.

A framework for testing membrane controllers on real and simulated robots has also been developed (see figure 9.3). The framework integrates xml files which store robot behaviors in a platform-independent way, a simulator for numerical P systems [13] and Webots, a professional mobile robot simulator [5].

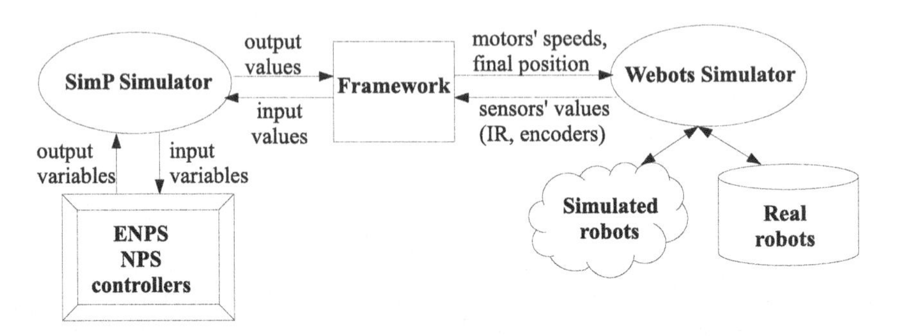

Fig. 9.3 Robotic framework

All developed membrane controllers have been tested both on real and simulated KheperaIII and e-puck robots (figure 9.4). Both types of robots have differential wheels and infrared sensors placed around the robot [9], [6].

(a) e-puck (b) KheperaIII

Fig. 9.4 NPS and ENPS controllers have been tested on real and simulated e-puck and KheperaIII robots

Membrane controllers modeled using ENPS have less complicated structures than the ones modeled with NPS. For instance, the ENPS controller for obstacle avoidance proposed in [14] is a membrane system with 9 membranes, while the NPS model for obstacle avoidance proposed in [2] has 37 membranes, as it can be noticed in figure 9.5. Detailed explanations regarding the the variables and the rules can be found in the referenced articles. Although the ENPS model for obstacle avoidance has more rules than the NPS model, not all of them are active (fewer rules are executed during a computational step in the ENPS model than in the NPS one). The controllers take into account 8 infrared sensors of the robot. E-puck robot has 8 infrared sensors around the robot, while Khepera III has 9, but the information received from the back sensor is ignored. Sensors' placement of the robots is illustrated in figure 9.6.

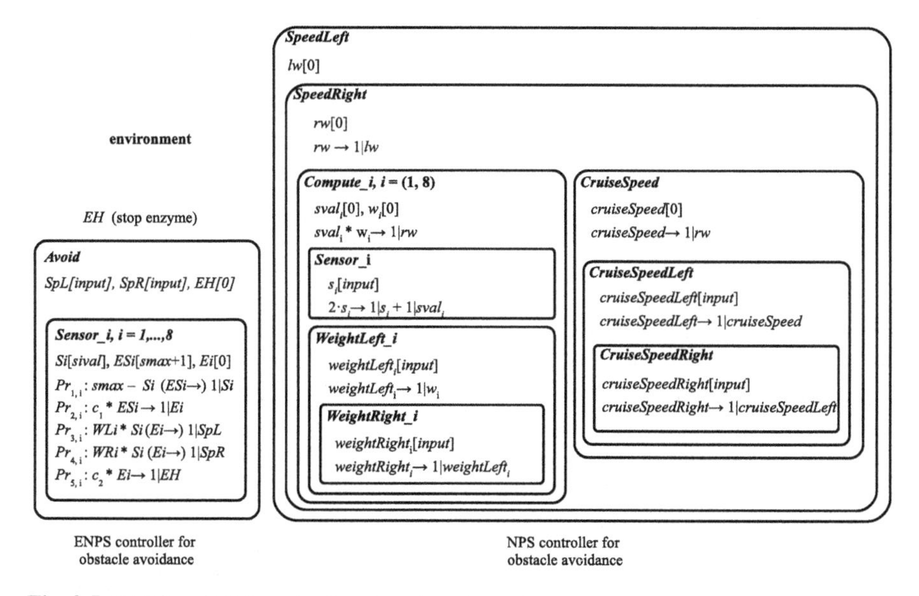

Fig. 9.5 ENPS and NPS controllers for obstacle avoidance

Avoiding obstacles is one of the main abilities that an autonomous robot must have in order to function safely and robust. Obstacle avoidance may be integrated as part of the robot's navigation system and is performed by setting the speeds of the robot's locomotion motors to appropriate values. The speeds of the motors (the actuators) must be controlled using the information from the sensors. Therefore, the membrane controllers compute the motors' speeds as weighted sums of the sensors' inputs. Both controllers implement the following control law:

$$SpeedLeft = cruiseSpeedLeft + \sum_{i=1}^{8} WeightLeft_i \cdot Sensor_i \tag{9.2}$$

$$SpeedRight = cruiseSpeedRight + \sum_{i=1}^{8} WeightRight_i \cdot Sensor_i \tag{9.3}$$

Another important problem in robotics is localization. The robot should be able to compute it's position at any moment based on the previous values. Therefore, a

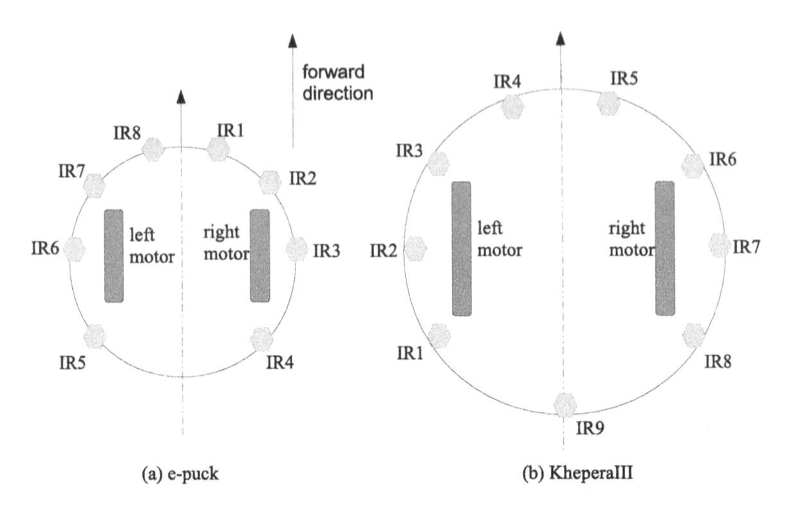

Fig. 9.6 Sensors' placement

NPS structure for odometric localization has been modeled as well, but its structure is far more complicated than the ENPS one proposed in [16]. In this case, the NPS controller was modeled using 24 membranes, while the ENPS model has only 5 membranes. In this case, enzymes control the program flow and are used as stop conditions and synchronization mechanism. Therefore, the model of the controller is clearly simplified, easier to implement and more efficient (regarding the computational process) than the one modeled by classical NPS.

The naturally parallelized membrane representation and the numerical nature of the membrane components represent advantages for both NPS and ENPS in designing and modeling robot behaviors.

Based on the theoretical and practical results, some important advantages of using ENPS to classical NPS are mentioned in [14] and [16]. The main advantages of ENPS towards NPS are that enzyme-like variables can control the program flow by deciding which rules to be executed in a computational step, they can control the synchronization between parallel computations, they can be used to filter noise from the sensors or to detect the termination of the program.

A comparison between a membrane controller (NPS) and a fuzzy controller for obstacle avoidance is presented in [2]. The authors show that a Mamdani type fuzzy logic controller with three inputs, each having 5 Gaussian membership functions and a rule base of 125 rules, has poorer performance than a NPS membrane controller. The fuzzy controller avoids obstacles very sharply while the NPS controller avoids obstacle more smoothly. The ENPS controller for obstacle avoidance has a simpler structure and a greater performance than the NPS controller as it was previously mentioned. Therefore, ENPS membrane controllers can be used for obstacle avoidance with comparable or better performance than fuzzy logic controllers.

9.4 Conclusions and Future Developments

This paper presents the currently existing results in modeling robot controllers by means of P systems. Numerical P systems and its extension, enzymatic numerical P systems, have been used to achieve cognitive robot behaviors.

Taking into account the most important properties of numerical P systems: their numerical nature, parallelization, distribution and the tree-like structure of the membranes, a future aim of this research direction is to design hardware membrane controllers and prove their efficiency (for example implementing the membrane controllers on a FPGA which can be connected to the robot).

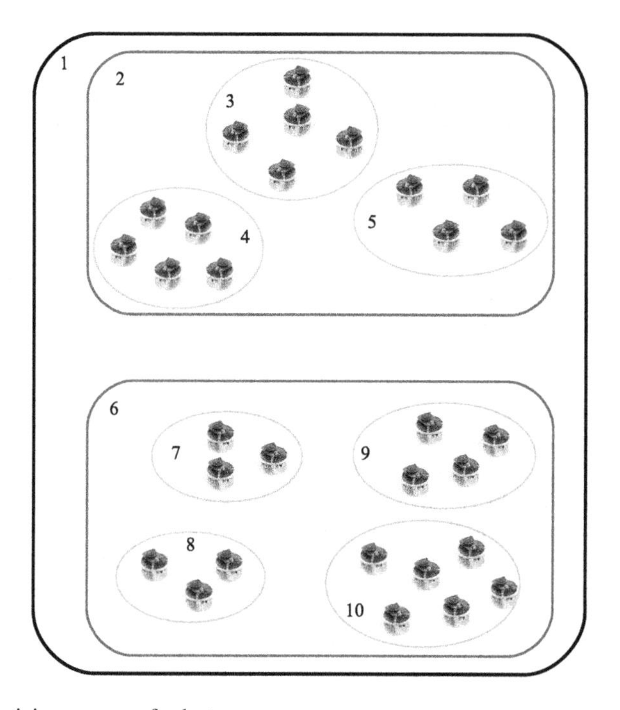

Fig. 9.7 Organizing groups of robots

A parallelized simulator for ENPS, implemented on GPU using CUDA (Compute Unifed Device Architecture) programming model, is proposed in [7]. Future developments include ENPS models for controlling a swarms of robots. For simulating more behaviors at the same time, the performance of the whole system would be much improved by using the ENPS GPU simulator proposed in [7]. Other robot behaviors can be modeled by using ENPS and a library of ENPS models for robotics applications is currently being developed. In this way the membrane controllers can be reused and stored as modules independent of the robotics system's software. For instance, membrane controllers can be integrated in Chidori multi-agent architecture [20]. The system is composed of three types of agents, a graphical user interface agent, and a pair of a local and a social agent for each robot in the swarm.

A hybrid system which uses both numerical P systems and symbolical P systems is an idea which the authors find interesting to explore. The local behaviors of the agents can be modeled using ENPS, while the social behavior and the interraction between robots can be modeled using symbolical P systems. For example the robots may form subswarms in order to devide tasks (figure 9.7). A membrane system can be used to organize robots in groups which are able to communicate and exchange information.

Using membrane systems in robotics is a new approach. Based on the results obtained so far, the authors consider further investigation on numerical P systems will contribute to the advancement of the robotics research.

Acknowledgements. This paper is supported by the Sectorial Operational Programme Human Resources Development, financed from the European Social Fund and by the Romanian Government under the contract number SOP HRD/107/1.5/S/82514.

References

1. Arsene, O., Buiu, C., Popescu, N.: SNUPS – A simulator for numerical membrane computing. Intern. J. of Innovative Computing, Information and Control 7(6), 3509–3522 (2011)
2. Buiu, C., Vasile, C., Arsene, O.: Development of membrane controllers for mobile robots. Information Sciences 187, 33–51 (2012), doi:10.1016/j.ins.2011.10.007
3. Cardona, M., Colomer, M.A., Pérez-Jiménez, M.J., Sanuy, D., Margalida, A.: Modeling Ecosystems Using P Systems: The Bearded Vulture, a Case Study. In: Corne, D.W., Frisco, P., Păun, G., Rozenberg, G., Salomaa, A. (eds.) WMC9 2008. LNCS, vol. 5391, pp. 137–156. Springer, Heidelberg (2009)
4. Colomer, M.A., Lavín, S., Marco, I., Margalida, A., Pérez-Hurtado, I., Pérez-Jiménez, M.J., Sanuy, D., Serrano, E., Valencia-Cabrera, L.: Modeling Population Growth of Pyrenean Chamois (Rupicapra p. pyrenaica) by Using P-Systems. In: Gheorghe, M., Hinze, T., Păun, G., Rozenberg, G., Salomaa, A. (eds.) CMC 2010. LNCS, vol. 6501, pp. 144–159. Springer, Heidelberg (2010)
5. Cyberbotics, professional mobile robot simulation (2012), `http://www.cyberbotics.com`
6. e-puck robot website (2012), `http://www.e-puck.org`
7. Garcìa-Quismondo, M., Pèrez-Jimènez, M.J.: Implementing ENPS by Means of GPUs for AI Applictions. In: Proceedings of Beyond AI: Interdisciplinary Aspects of Artificial Intelligence (BAI 2011), Pilsen, Czech Republic, pp. 27–33 (2011)
8. Hoffmann, H., Maggio, M., Santambrogio, M.D., Leva, A., Agarwal, A.: SEEC: A General and Extensible Framework for Self-Aware Computing. Technical Report MIT-CSAIL-TR-2011-046 (2011)
9. Lambercy, F., Caprari, G.: Khepera III manual ver 2.2 (2012), `http://ftp.k-team.com/KheperaIII/UserManual/Kh3.Robot.UserManual.pdf`
10. Manca, V., Marchetti, L.: Log-Gain Stoichiometric Stepwise Regression for MP Systems. International Journal of Foundations of Computer Science 22(1), 97–106 (2011)

11. Manca, V., Marchetti, L.: Solving dynamical inverse problems by means of Metabolic P systems. Bio Systems (in press)
12. Marchetti, L., Manca, V.: A Methodology Based on MP Theory for Gene Expression Analysis. In: Gheorghe, M., Păun, G., Rozenberg, G., Salomaa, A., Verlan, S. (eds.) CMC 2011. LNCS, vol. 7184, pp. 300–313. Springer, Heidelberg (2012)
13. Pavel, A.B.: Membrane controllers for cognitive robots. Master Thesis, Department of Automatic Control and System Engineering, Politehnica University of Bucharest (2011)
14. Pavel, A.B., Buiu, C.: Using enzymatic numerical P systems for modeling mobile robot controllers. Natural Computing (2011), doi:10.1007/s11047-011-9286-5
15. Pavel, A.B., Arsene, O., Buiu, C.: Enzymatic numerical P systems – a new class of membrane computing systems. In: The IEEE Fifth Intern. Conf. on Bio-Inspired Computing. Theory and applications (BIC-TA 2010), Liverpool, U.K., pp. 1331–1336 (2010)
16. Pavel, A.B., Vasile, C., Dumitrache, I.: Robot localization implemented with enzymatic numerical P systems (submitted)
17. Păun, G.: Computing with membranes. Journal of Computer and System Sciences 61, 108–143 (2000)
18. Păun, G., Rozenberg, G., Salomaa, A. (eds.): The Oxford Handbook of Membrane Computing. Oxford University Press (2010)
19. Păun, G., Păun, R.: Membrane computing and economics: Numerical P systems. Fundamenta Informaticae 73(1), 213–227 (2006)
20. Vasile, C., Pavel, A., Buiu, C.: Integrating human swarm interaction in a distributed robotic control system. In: IEEE 7th Annual IEEE Conference on Automation Science and Engineering, Trieste, Italy (2011)
21. Vasile, C., Pavel, A.B., Dumitrache, I., Păun, G.: On the Power of Enzymatic Numerical P Systems (submitted)

Chapter 10
Implementing Enzymatic Numerical P Systems for AI Applications by Means of Graphic Processing Units

Manuel García–Quismondo, Luis F. Macías–Ramos, and Mario J. Pérez–Jiménez

Abstract. A P system represents a distributed and parallel computing model in which basic data structures are, for instance, multisets and strings. Enzymatic Numerical P Systems (ENPS) are a type of P systems whose basic data structures are sets of numerical variables. Separately, GPGPU (general-purpose computing on graphics processing units) is a novel technological paradigm which focuses on the development of tools for graphic cards to solve general purpose problems. This paper proposes an ENPS simulator based on GPUs and presents general concepts about its design and some future ideas and perspectives.

10.1 Introduction

Membrane computing is a bio-inspired branch of natural computing, abstracting computing models from the structure and functioning of living cells and from the organization of cells in tissues or other higher order structures [29]. This branch of natural computing studies the design and properties of membrane systems or *P systems*. P systems are non–deterministic distributed and parallel computing models structured in compartments known as *membranes*. Basic data structures such as multisets, strings or numerical variables [30] are associated with membranes. According to the way in which membranes are structured, there are several types of P systems. For instance, there exist cell-like P systems [29], tissue P systems [26] and spiking neural P systems [17], along with other types. In P systems, membranes and their associated data structures are processed by means of rewriting rules or programs associated to the cells, in order to perform sequences of configurations (*computations*) [29][30]. P systems have been successfully applied in a wide range of domains [4]. For instance, they have been applied in microbiological modelling

Manuel García–Quismondo · Luis F. Macías–Ramos · Mario J. Pérez–Jiménez
Research Group on Natural Computing, Dpt. of Computer Science and Artificial Intelligence, University of Sevilla, Avda. Reina Mercedes s/n. 41012 Sevilla, Spain
e-mail: {mgarciaquismondo,lfmaciasr,marper}@us.es

J. Kelemen et al. (Eds.): Beyond Artificial Intelligence, TIEI 4, pp. 137–159.
DOI: 10.1007/978-3-642-34422-0_10 © Springer-Verlag Berlin Heidelberg 2013

in order to model phenomena such as *quorum sensing* in *Vibrio fischeri* populations [38] and ecological modelling to predict the evolution of the bearded vulture [3] and the Pyrenean chamois [9] populations in the Catalan Pyrenees, as well as image thresholding [7]. Such a versatility makes P systems a useful tool for gaining knowledge about a vast variety of different domains, thus providing a promising tool whithin the range of disciplines which composes the field of study of artificial intelligence.

A special type or cell-like P systems are Enzimatic Numerical P Systems (ENPSs) [36]. ENPSs describe a deterministic, maximally–parallel model in which the basic data structures associated to membranes are numerical values which evolve by means of *programs* associated to them [30]. In order for a program to be applied, a certain value of a specific variables (enzyme) may be needed. Otherwhise, the program cannot be applied [36]. This model of computation has already been successfully used in model robot controllers, in which a robot needs to avoid obstacles situated in a closed circuit [37].

Separately, *GPGPU* (general-purpose computing on graphics processing units) is a novel technological discipline which consists of the application of graphic cards (GPUs) in order to perform parallel, distributed algorithms [40]. The basic idea is to take advantage of the parallel architecture of GPUs, traditionally used for graphics processing, to execute algorithms which can be performed in parallel, thus accelerating these algorithms by dividing them in concurrent tasks and executing these tasks in a parallel mode.

In this paper, we propose a GPU simulator for ENPSs. The parallel architecture of ENPS makes the simulations of their computations a suitable task to be parallelized, thus expecting an acceleration in the simulation times if compared to their sequential counterparts.

This paper is structured as follows. Section 10.2.2 provides a quick introduction to Numerical P Systems (NPSs) as a model of computation. Section 10.3 describes ENPSs as an extension of NPSs. Section 10.4 provides a general overview of the current state-of-the-art about the results obtained by previous GPU simulators within the field of membrane computing. Finally, section 10.7 presents the conclusions obtained and proposes some directions for future work.

10.2 Preliminaries

10.2.1 P Systems

Membrane Computing is a young and emergent branch of Natural Computing introduced by G. Păun [30]. It has received important attention from the scientific community since then, with contributions by computer scientists, biologists, formal linguists and complexity theoreticians, enriching each others with results, open problems and promising new research lines. In fact, membrane computing was selected by the Institute for Scientific Information, USA, as a fast *Emerging Research Front* in computer science, and [35] was mentioned in [42] as a highly cited paper in October 2003. This new model of computation starts from the observation that the cell is the smallest living thing, and at the same time it is a marvellous tiny machinery, with a complex structure, and from the assumption that the processes taking place in the compartmental structure of a living cell can be interpreted as computations. The challenge is to take the cell itself as a support for computations, to find in the structure and the functioning of the cell seen as a whole those elements useful for computations. Computations in general, at the mathematical level, but with the hope to bring something useful to practical computing, either in the same style as genetic algorithms and neural computing, of improving the use of the existing computers, or proposing new types of electronic computers, or, possibly to lead to ways to use the cells themselves as computing supports. The devices of this model are called P systems. Roughly speaking, a P system consists of a cell-like membrane structure, in the compartments of which one places multisets of objects which evolve according to given rules.

The main syntactic ingredients of a cell-like membrane system are the membrane structure, the multisets of objects, and the evolution rules. A membrane structure consists of several membranes arranged in a hierarchical structure inside a main membrane (the skin), and delimiting regions (the space in–between a membrane and the immediately inner membranes, if any). Each membrane identifies a region inside the system Regions defined by a membrane structure contain objects corresponding to chemical substances present in the compartments of a cell. The objects can be described by symbols or by strings of symbols, in such a way that multiset of objects are placed in regions of the membrane structure. The objects can evolve according to given evolution rules, associated with the regions (hence, with the membranes).

The semantics of the cell-like membrane systems is defined through a non deterministic and synchronous model (in the sense that a global clock is assumed) as follows: A configuration of a cell–like membrane system consists of a membrane structure and a family of multisets of objects associated with each region of the structure. At the beginning, there is a configuration called the initial configuration of the system. In each time unit we can transform a given configuration in another configuration by applying the evolution rules to the objects placed inside the regions of the configurations, in a non-deterministic, and maximally parallel manner (the rules are chosen in a non-deterministic way, and in each region all objects that can evolve must do it). In this way, we get transitions from one configuration of the system to the next one.

In the last years, many different models of P systems have been proposed. In particular, computational devices inspired from the cell inter–communication in tissues, and adding the ingredient of cell division rules of the same form as in cell–like membrane systems with active membranes, but without using polarizations. In these systems, the rules are used in the non-deterministic maximally parallel way, but we suppose that when a cell is divided, its interaction with other cells or with the environment is blocked; that is, if a division rule is used for dividing a cell, then this cell does not participate in any other rule, for division or communication. The set of communication rules implicitely provides the graph associated with the system through the labels of the membranes. The cells obtained by division have the same labels as the mother cell, hence the rules to be used for evolving them or their objects are inherited.

The idea of spiking neurons, currently an active research topic in neural computing (see, e.g., [16], [22], [23]), was recently incorporated in membrane computing (see [18]) – the resulting formal systems are called *spiking neural P systems*, abbreviated as SN P systems. The structure of an SN P system has a form of a directed graph with nodes representing neurons, and edges representing synapses. The neurons contain *spikes*, objects of a unique type. A neuron (node) sends signals (spikes) along its outgoing synapses (edges). Each neuron has its own rules for either sending spikes (firing rules) or for internally consuming spikes (forgetting rules). the rules of the first type consume some spikes and produce a new spike, which is sent to all neurons linked by a synapse to the neuron where the rule was used, while the forgetting rules just remove spikes from neurons. In the initial configuration a neuron stores the initial number of spikes, and at any time moment the currently stored number of spikes (*current contents*) is determined by the initial contents and the history of functioning of σ (the spikes it received from other neurons, the spikes it sent out, and the spikes it internally consumed/forgot). One of the neurons is the *output* one, and its spikes can also exit into the environment, thus providing a trace of the system evolution. Like in neurobiology, we call this trace – sequence of moments when a spike exits the system – *spike train*.

10.2.2 Numerical P Systems

As years went by, different types of P systems have been introduced. In the foundational transition P system model, the cell structure consists of a rooted tree, in which each node represents a membrane of the structure. Edges represent the hierarchical relationships between membranes existent in the structure. However, some models propose new types of cell structures. For instance, SN P Systems describe an architecture based on a directed graph, in which cells or *neurons* act as nodes, whereas firing rules act as arcs. These rules send information from one neuron to another after a specific amount of time or delay [19]. Similarly, in Tissue P systems, instead of a hierarchical structure, membranes are placed at the nodes of a non-directed graph. The edges of the graph represent symport/antiport rules which communicate the membranes in the graph, thus moving objects across membranes [26]. Also, even variants of these ones have evolved. For instance, in the case of SN P Systems, new features such as SN P Systems with several kinds of spikes [19] and SN P systems with neuron division and budding [27]. As regards to Tissue P Systems, there exist Tissue P Systems with cell division [33], Tissue P Systems without environment [8] as an example.

Besides, not only have membrane structures evolved across the Membrane Computing literature. The data structures which evolve by means of applications of rules through computation steps have also been affected. As a proof of that, in String P Systems sets of strings are considered instead of multisets of objects. These strings are rewritten by means of rewriting–like rules on each computation step [6].

Following this trend, a new kind of P system was introduced by Gheorghe and Andrei Păun in 2006. In these P systems, known as Numerical P Systems (*NPSs* [29]), the traditional multisets of objects associated to membranes are replaced by sets of numerical variables. These variables evolve by means of programs associated to the membranes. As in the foundational model, the membrane structure is a tree-nested hierarchy, so no new membrane architecture is introduced in this model.

A numerical P system of degree $m \geq 1$ is a tuple:

$$\Pi = (H, \mu, (Var_1, Pr_1, Var_1(0)) \ldots (Var_m, Pr_m, Var_m(0)))$$

where:

- H is an **alphabet** with m symbols used as labels of the m membranes of the system. The labels contained in H are the labels of the membranes in Π.
- μ is a **membrane structure**, a rooted tree with m membranes.
- $Var_i = \{x_{1,i} \ldots x_{k_i,i}\}$ is the set finite of **variables** associated with compartment i, $(1 \leq i \leq m)$
- $Var_i(0) = \{\lambda_{1,i} \ldots \lambda_{k_i,i}\}$ are numerical values (*real numbers*) for the variables in Var_i. These values are considered as initial values; at time = of the system evolution we have $x_{j,i} = \lambda j, i(1 \leq i \leq m, 1 \leq j \leq k_i)$.
- $Pr_i = Pr_{1,i} \ldots Pr_{q_i,i}$ is the set of programs from comparment i of $\mu(1 \leq i \leq m)$. The *l-th* program $Pr_{l,i}$ from compartment i is of the form $Pr_{l,i} = (F_{l,i}(x_{1,i}, \ldots, x_{k_i,i}), c_{l,1}|v_1 + \ldots + c_{l,n_i}|v_{n_i})$ where $F_{l,i}(x_{1,i}, \ldots, x_{k_i,i})$ is the *l-th* **production function** from compartment i and $c_{l,1}|v_1 + \ldots + c_{l,n_i}|v_{n_i})$ describes the **repartition protocol**.

The production function $F_{l,i}(x_{1,i}, \ldots, x_{k_i,i})$ from compartment i is a a real function having as variables those from this compartment. The expresion $c_{l,1}|v_1 + \ldots + c_{l,n_i}|v_{n_i}$ describes the repartition protocol which has the following meaning: let $v_1 \ldots v_{n_i}$ be the set of variables from compartment i, from the parent membrane of i and for all compartments corresponding to children of comparment i. The coefficients $c_{l,1} \ldots + c_{l,n_i}$ are natural numbers that specify the proportion of tehe current production distributed to each variable $v_1 \ldots v_{n_i}$.

More precisely, at any instant $t \geq 0$, a program $Pr_{l,i}$ on each set Pr_i $(1 \leq i \leq m)$ is non–deterministically chosen. Then, we compute $F_{l,i}(x_{1,i}(t), \ldots, x_{k_i,i}(t))$ and $C_{l,i} = \sum_{j=1}^{n_i} c_{l,j}$. The values of all variables on which $F_{l,i}$ depends are *consumed* and reset to 0. The value $q = \frac{F_{l,i}(x_{1,i}(t), \ldots, x_{k_i,i}(t))}{C_{l,i}}$ represents the "unitary portion" to be distributed to variables $v_1, \ldots, v_{n_i}$, according to coefficients $c_{l,i}, \ldots, c_{l,n_i}$ in order to obtain the values of these variables at time $t+1$. Specifically, variable $v_{l,j}$ will receive $q \times c_{l,j}(1 \leq j \leq n_i)$ from compartment i. If a variable receives such "contributions" from several neighbouring compartments, then they are added in order to produce the value of the variable at time $t+1$.

This model of computation was initially aimed to capture the nature and behaviour of economic processes [29]. There had been some previous works on the modelling of economic processes by means of Membrane Computing [34], and this work proposed some research lines on the application of NPSs for the modelling of economic phenomena.

10.3 Enzymatic Numerical P Systems

10.3.1 Description of Enzymatic Numerical P Systems

As it is usual on membrane computing models, a new kind of P systems has risen as an extension of NPSs. This model is known as Enzymatic Numerical P Systems (*ENPSs*). Although this parallel model of computation has many points in common with Numerical P Systems, there are some aspects which differenciates both models. This way, in contrast to Numerical P Systems, Enzymatic Numerical P Systems describe a deterministic model of computation. Thus, instead of non–deterministically chosen, the programs to be applied are controlled by specific variables known as *enzyme–like* variables.

An Enzymatic Numerical P System of degree $m \geq 1$ is a tuple:

$$\Pi = (H, \mu, (Var_1, Pr_1, Var_1(0)) \dots (Var_m, Pr_m, Var_m(0)))$$

where:

- H, μ and $(Var_1, Var_1(0)) \dots (Var_m, Var_m(0))$ have the same meaning than in Numerical P Systems described in section 10.2.2.
- Pr_i is the set of programs associated to membrane i. Each l-th program in set Pr_i may have one of the following forms:
  - $Pr_{l,i} = (F_{l,i}(x_{1,i}, \dots, x_{k_i,i}), c_{l,1}|v_1 + \dots + c_{l,n_i}|v_{n_i})$
  - $Pr_{l,i} = (F_{l,i}(x_{1,i}, \dots, x_{k_i,i}), (e_{l,i} \rightarrow), c_{l,1}|v_1 + \dots + c_{l,n_i}|v_{n_i})$

  In both forms, all values which also appear in section 10.2.2 have the same meaning, with $e_{l,i}$ being a variable in Var_i. This variable is known as the *enzyme–like* variable associated to $Pr_{l,i}$ and its value cannot be consumed by this program. Enzyme–like variables are exclusive ingredients of ENPSs. That is, they do not appear in NPSs.

The main novelty introduced by ENPSs has to do with the use of enzyme–like variables to control the execution flow of programs. This way, each program may have an associated enzyme–like variable which controls its application. If a program is to be applied at time t, then this program is *active* at this time. On each computation step, all active programs in each membrane are applied in parallel. Programs in ENPSs are applied the same way than in NPSs. However, a program is active only in the following cases:

- The program does not have an associated enzyme.
- The program has an associated enzyme and the value of this enzyme is greater than the minimum of the values of the variables consumed by the program.

ENPSs have been successfully applied within the field of robotics. For instance, they have been used to model deterministic mobile robot controllers for obstacle avoidance. In this model, the speed of the two robot motors is set according to the values assigned to two variables of the system. Thus, the dynamical evolution of these variables describes the behavior of the robot through a closed circuit [37]. More information about ENPSs can be found in [36][37].

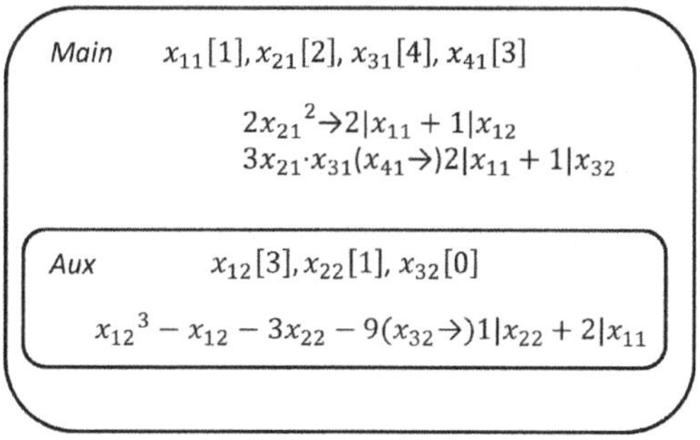

Main $x_{11}[1], x_{21}[2], x_{31}[4], x_{41}[3]$

$$2x_{21}{}^{2} \rightarrow 2|x_{11} + 1|x_{12}$$
$$3x_{21} \cdot x_{31}(x_{41} \rightarrow) 2|x_{11} + 1|x_{32}$$

Aux $x_{12}[3], x_{22}[1], x_{32}[0]$

$$x_{12}{}^{3} - x_{12} - 3x_{22} - 9(x_{32} \rightarrow) 1|x_{22} + 2|x_{11}$$

Fig. 10.1 Enzymatic Numerical P System

10.3.2 ENPSs and Artificial Intelligence

Mobile robot control problems, such as obstacle avoidance and odometric localizacion, can be considered as artificial intelligence problems. For instance, obstacle avoidance can be considered as a high-level planning problem [21]. In obstacle avoidance, the objective is to find a sequence of movements in a static or dynamical environment. The objective of this sequence is for robots which follow it to avoid crashing with any obstacles they might find in the environment. The input data is given as a series of sensor lectures obtained from the environment. This type of path planning problems arising from the field of robotics has already been attacked by using artificial intelligence techniques such as ant colony algorithms [12][11].

Odometric localization is a widely used method for estimation of the momentary pose of a mobile robot with respect to its starting pose [20]. This estimation is affected by several error sources, such as imprecission in the mobile robot kinematic parameters and errors in the sensor lectures [1]. Thus, odometric localization entails an optimization problem, i.e., minimizing the global error in the pose estimation. As an optimization problem, odometric localization has been previously tackled by using well-known artificial intelligence paradigms, such as genetic algorithms [15] and artificial neural networks [10]. All in all, ENPs propose a new framework which can be applied in order to solve artificial intelligence problems arising from robotics [37].

10.3.3 Simulation of ENPSs

ENPSs describe a parallel model. Therefore, the huge computational power required by extensive models (for instance, those necessary for massive robot swarms and robots with complex sensor networks) accounts for the need for high performance computing platforms to simulate them. Besides, their parallel structure makes them appropriate to be simulated by means of parallel architectures such as GPUs, FPGAs and computer clusters.

10.4 The Compute Unified Device Architecture (CUDA) Standard for GPU Computing

10.4.1 Outline of the CUDA Programming Model

Modern GPUs consist of a large number or processing units. For instance, Fermi cards contain up to 448 processor cores and 1.536 processing units per core, thus resulting in a total number of $448 \times 1.536 = 688.128$ threads [41]. These threads are executed in parallel with a certain degree of dependency from each other [40].

In order to make the most of this massively parallel architecture, it is necessary to make use of standards specifically designed for these devices. Two of these main standards in GPGPU are OpenCL [39] and CUDA [41].

The CUDA programming model is an abstract GPU model provided by NVIDIA. This model is an abstraction of the specific parallel device where the program is to be executed. The model defines a *grid*. This grid is an abstraction of the current GPU card where the code is to be executed. The grid is composed of multiprocessing computing devices known as *blocks*. Similarly, each block is composed of several stream monoprocessing units known as *threads*(see figure 10.2). Threads execute parallel pieces of code or *kernels*. On any instant in the execution of a GPU program, the same kernel is run on every thread at the same time.

It is convenient to batch threads which perform operations in common in the same block. The reason is that threads in the same block can communicate with each other through fast on-chip memory, whereas threads in different blocks use slow off-chip memory to communicate. Thus, it is important to minimize the communication between threads from different blocks, turning it into communication between threads in the same block when possible. Besides, they are allowed to synchronize with each other via barriers. On the other hand, the only way of synchronizing threads of different blocks is by ending the kernel execution. The CUDA programming model requires thread blocks in the same kernel to be independent. It means that the final result of the computation cannot depend on the order in which the blocks are executed, giving the same result without depending on their order of execution.

10.4.2 The CUDA–C *Programming Language*

CUDA–C is an extension of the C language to work against the CUDA programming model. This language is designed to make the most of the GPGPU approach by enabling programmers to encode parallel applications to be run on GPUs [5]. That is, programmers are able to develop code to be executed on each GPU thread at the same time. This way they can take advantage of the GPU parallel architecture in order to obtain enormous speed-up if compared to sequential versions of the same code.

The structure of CUDA-C programs consists of two main parts: The *host* part and the *device* part. The main difference between them consists of the specific device in which they are executed. Thus, the host part is executed on the CPU, whilst the device part is executed on the GPU [5]. The host part includes calls to kernels. The

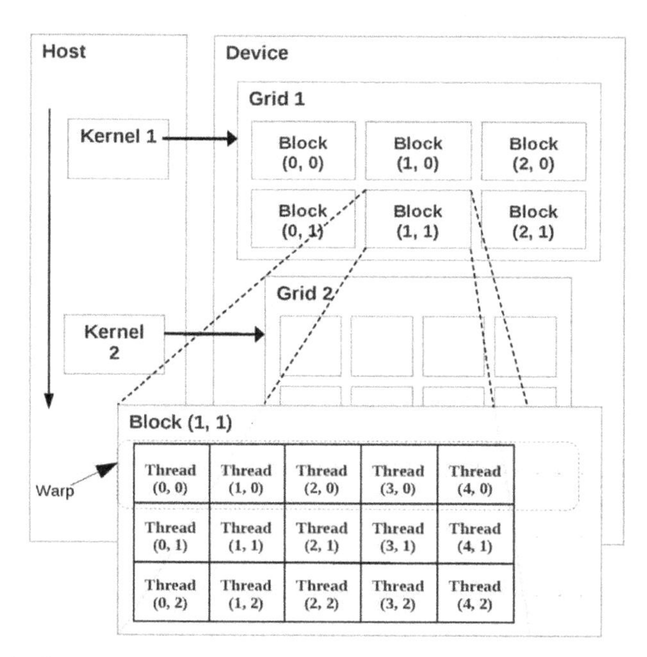

Fig. 10.2 The CUDA programming model

device part is composed of kernels which define the operations to be performed in parallel. The developer organizes the threads to execute the kernels in two hierarchical levels of parallelism. These levels are a reflection of those in which the CUDA programming model is organized. In order the organize the threads to execute the kernels, the programmer defines the structure of the thread blocks. This is done by programmatically setting the number of threads per block, as well as the total number of blocks in the grid. This way, both parts of the program can cooperate in order to obtain a global result. More information about the CUDA programming model and the CUDA-C language can be found on [41][24].

A sample code of a typical high-performance operation on GPU can be found on [4]. In this sample, the summing of the elements in two vectors is computed. Each pair of elements are assigned to a different thread. Therefore, each pair of elements are added in a parallel way. Although this example may seem too simple, it illustrates quite well the way in which the CUDA parallel mode can be applied to parallelize operations, thus obtaining a tremendous speed-up due to the parallel computing approach.

GPGPU and CUDA–C have been already successfully applied in order to simulate different kinds of P systems. To the best of our knowledge, they have been applied to simulate cell-like object-based P systems [5] and SN P systems [2]. Their results include data which show noticeable speed-ups in comparison to their sequential counterparts. These results demonstrate the suitability of the GPGPU approach for simulating P systems in a parallel mode.

```
//Vector size in elements
const int N = 1048576;
//Vector size in bytes
const int dataSize = N * sizeof(float);

//CPU memory allocation
float *h_A = (float *)malloc(dataSize);
float *h_B = (float *)malloc(dataSize);
float *h_C = (float *)malloc(dataSize);

//GPU memory allocation
float *d_A, *d_B, *d_C;
cudaMalloc((void **)&d_A, dataSize));
cudaMalloc((void **)&d_B, dataSize));
cudaMalloc((void **)&d_C, dataSize));

//Initialize h_A[], h_B[]

//Copy input data to GPU for processing
cudaMemcpy(d_A, h_A, dataSize, cudaMemcpyHostToDevice) );
cudaMemcpy(d_B, h_B, dataSize, cudaMemcpyHostToDevice) );

//Run the core of N / 256 units. 256 streams each
//Assuming that N is multiple of 256
vectorAdd<<<N / 256, 256>>>(d_C, d_A, d_B);

//Read GPU results
cudaMemcpy(h_C, d_C, dataSize, cudaMemcpyDeviceToHost) );
```

Fig. 10.3 A sample of CUDA–C host code

10.5 A GPU–Based Simulator for Enzymatic Numerical P Systems

Taking into account these previous results, we propose a new simulator for ENPSs developed on CUDA-C. Thus, we expect our simulator to achieve an important acceleration of the execution times in comparison to currently existent ENPS sequential simulators [36].

In this section, the guidelines for the design of the current version of the simulator are outlined. Moreover, a thorough description of the data structures and the functioning of the simulator is explained.

The objective of the proposed ENPS GPU-based simulator is to fully simulate the behaviour of enzymatic numerical P systems, performing operations in parallel whenever possible. In order to do that, it is crucial to identify which operations are susceptible for parallelization and write parallel kernels for them. This way the simulator can take advantage of the underlying parallel architecture.

10.5.1 Data Representation

As it is usual in GPU computing [4], the data handled by the simulator is stored by means of arrays. The simulator uses three different kinds of arrays, according to the nature of the information stored in them:

Program arrays: These arrays are used to store the programs of the ENPS model simulated. These arrays can be organized in three different types:

- **Production function:** These arrays are used to store the information regarding the production functions of rules. They are described in subsection 10.5.4.1 in detail.
- **Repartition protocol:** These arrays are used to store the information about the repartition protocols. They are described in subsection 10.5.6 in detail.
- **Enzymes:** Each program in the simulated ENPS model has an associated position in this array. This position contains the index of the enzymatic variable

associated to the program. In the case that the program is in non-enzymatic form, the position contains a specific marker, such as -1.

Variables: This array stores the value of the variables associated to the compartments in the ENPS model simulated. These values evolve as programs in the model are applied.

Auxiliary data: These arrays store the auxiliary data needed in order to check and apply the programs. Specifically, these arrays are:

Minimum values: This array stores the minimum values of the variables consumed by programs

Production function results: This array stores the results of the calculations of the applied production functions

Program applications: This array stores themarkers to set if programs are applied. These markers can be *Active* or *Inactive*.

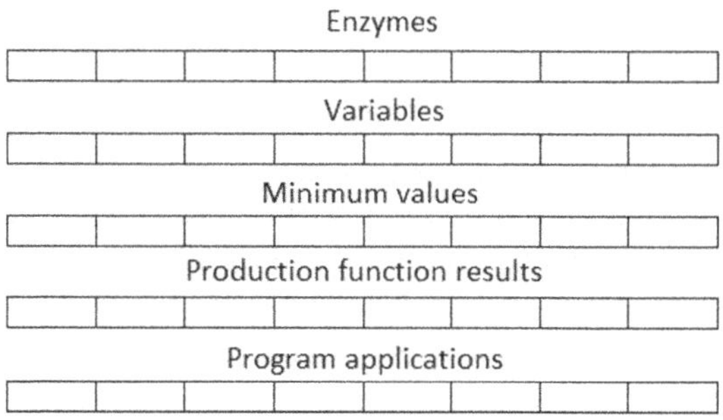

Enzymes

Variables

Minimum values

Production function results

Program applications

Fig. 10.4 Arrays used by the simulator to store general information

10.5.2 Repartition Coefficients Normalization

This step is only taken once, as it is a pre–processing operation in order to improve the efficiency of the simulator. It is not performed in parallel, thus being part of the *host* code. For each repartition protocol in each program $P_{l,i}$, each coefficient $c_{l,s}$ associated to the repartition protocol in is replaced by $\frac{c_{l,s}}{\sum_{j=1}^{n_i} c_{l,j}}$ $(1 \leq s \leq n_i)$. Although the formal definition of both NPSs and ENPSs establishes that $c_{l,s} \in \mathbb{N}(1 \leq s \leq n_i)$, these new repartition coefficients are real numbers, as they are temporary values calculated in order to improve the efficiency of the simulation algorithm.

10.5.3 Program Checking

The first stage of the implemented algorithm consists of selecting which programs can be applied on the current step of computation. Therefore, for each program, one should distinguish two different cases:

- The program is in non-enzymatic form. This means that no enzyme–like variable is associated to the program. In this case, the program is always executed. In such a case, the program's associated position in *Enzymes* should be a specific marker, as shown in subsection 10.5.1. Hence, the program's associated position in *Program applications* is set to *Active*.
- The program is in enzymatic form. This means that an enzyme–like variable is associated to the program. In this case, the simulator needs to find the minimum value of the variables consumed by the program. Then, it is necessary to calculate the minimum value of all variables consumed by the program, in order to distinguish two different cases:
  - The value of this minimum is greater than or equal to the value of the associated enzyme–like variable. In this case, the program cannot be applied.
  - The value of this minimum is lower than the value of the associated enzyme-like variable. In this case, the program has to be applied.

  In terms of implementation, each thread has an associated index i in the production function arrays (see subsection 10.5.4 for more details). Thus, each thread has an associated program. Each thread whose associated program is in enzymatic form performs the following steps:

  - Step through the region of the arrays in subsection 10.5.4 associated to the production function and checking those positions in which the array *Production function node types* contains the value **variable**.
  - Check the value of the array *Production function variables* in these positions.
  - Use the value of these positions on each thread as indexes to access the array *Variables*.
  - Calculate the minimum of the positions in this array.
  - Compare this minimum to the value of the enzyme–like variable of its associated program.
    - If the value of the enzyme–like variable is greater, then the program's associated position in *Program applications* is set to *Active*.
    - If the value of the enzyme–like variable is lower or equal, the program's associated position in *Program applications* is set to *Inactive*.

10.5.4 Calculation of Production Functions

In this section, an outline of the performing of the calculation of production functions is presented. For doing so, firstly the data structures used to represent production functions are introduced. Secondly, the way in which these data structures are processed is described. Finally, a brief discussion about the expected speed–up factor ends this subsection.

10.5.4.1 Structural Design of Production Functions

In the presented simulator, production functions are represented as tree–like structures. In these tree–like structures there exist two different kinds of nodes:

Non-leaf nodes: These nodes represent binary operations. On these nodes, the operands could be constants, variables or the result of other operations.
Leaf nodes: These nodes represent constants or variables. In the case of variables, their value is the evaluation of its represented variable.

Production functions are implemented by means of five different arrays. Thus, each tree representing a production function is implemented as a region in these five arrays. Each node is implemented as a position in all these arrays. These arrays are described as follows:

Production function node types: For each node, this array denotes the type of the node. It could be *constant* or *variable* (leaf nodes) or anyone of the *operation* type (non-leaf nodes). Each node of the *operation* type tells the simulator to perform a binary operation on its children. The values for nodes of the *operation* type are:

Add: Add the value of the left child to the value of the right child.
Subtract: Subtract the value of the left child to the value of the right child.
Multiply: Multiply the value of the left child by the value of the right child.
Divide: Divide the value of the left child by the value of the right child.
Power: Power the value of the left child to the value of the right child.

Production function left offsets: Given a position in this array, if its corresponding position in *Production function node types* is equal to *constant* or *variable* then this position has no meaning. Otherwise, if its corresponding position in *Production function node types* is equal to anyone of the *operation* type, then this position contains the relative offset where the **left** operand of the represented node is stored.
Production function right offsets: Given a position in this array, if its corresponding position in *Production function node types* is equal to *constant* or *variable* then this position has no meaning. Otherwise, if its corresponding position in *Production function node types* is equal to anyone of the *operation* type, then this position contains the relative offset where the **right** operand of the represented node is stored.
Production function constants: Given a position in this array, if its corresponding position in *Production function node types* is equal to *variable* or anyone of the *operation* type then this position has no meaning. Otherwise, if its corresponding position in *Production function node types* is equal to **constant**, then this position contains the value of the constant represented by its node.
Production function variables: Given a position in this array, if its corresponding position in *Production function node types* is equal to *constant* or *operation* then this position has no meaning. Otherwise, if its corresponding position in

Production function node types is equal to **variable**, then this position contains the position in *Variables* of the variable represented by its node.

This way, one can associate an index i in all of these arrays to every node N in every production function tree–like structure.

Production function node types							

Production function left offsets							

Production function right offsets							

Production function variables							

Production function constants							

Fig. 10.5 Data structure for production functions

10.5.4.2 Functional Design of Production Functions

By making use of the data structures described above, calculating a production function can be simply reduced to stepping through the nodes of its representing tree. Then, the recursive algorithm used to calculate production functions is:

1. Given a node N, check the node type of N. In terms of implementation, it means checking its associated position i in *Production function node types* as listed above.
    a. If the node type of N is *constant*, then return the value of the constant associated to N. In terms of implementation, it means returning the value in position i in the array *Production function constants*.
    b. If the node type of N is *variable*, then return the value of the variable asociated to N. In terms of implementation, it means taking the value stored in position i in the array *Production function variables* and using this value j as an index to return position j in the array *Variables* described in section 10.5.1.
    c. If the node type of N is anyone of the *operator* type, then:
        i. Access position i in *Production function left offsets*. Let j be the content of this position.
        ii. Calculate the result of $i+j$. Let $k = i+j$.
        iii. Process the node N^l whose index is k. It means going back to step 1, but processing N^l instead of N.
        iv. Access position i in *Production function right offsets*. Let m be the content of this position.
        v. Calculate the result of $i+m$. Let $n = i+m$.
        vi. Process the node N^r whose index is n. It means going back to step 1, but processing N^r instead of N.

vii. Apply the operation indicated by position i in *Production function operators* to the result of processing N^l as left child and the result of processing N^r as right child.
viii. Return the result of this operation.
d. Store the result of the calculation in the program's associated position o in *Production function results*.

This algorithm is executed by each of the threads of the kernels, if and only if their associated program is active. Thus, in this version of the simulator the theoretical speed–up factor on the calculation of production functions is equal to the number of programs of the simulated model. One could argue that some operations in these tree steppings could be performed in parallel, thus improving the theoretical speed–up factor. For, instance, in the production function represented in figure 10.6, $x_{1,2}+7$ and $x_{1,4}-3$ can be performed in parallel.

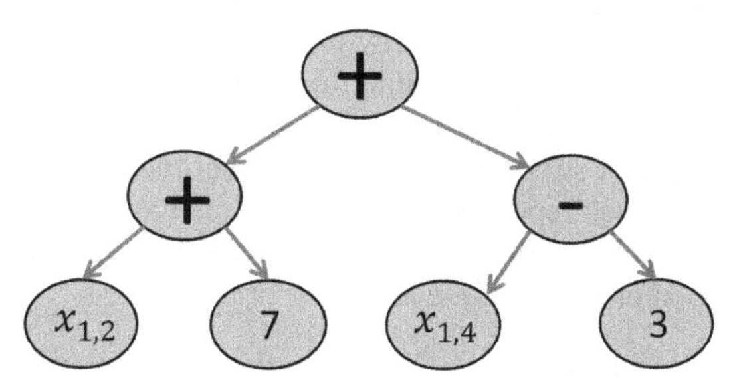

Fig. 10.6 Tree representing the production function $(x_{1,2}+7)+(x_{1,4}-3)$

However, the vast variety of different cases which can be found in these complex production functions makes it really difficult to design an efficient parallel implementation of their binary operations. Besides, production functions are usually short (though sometimes complex) functions [36], so the speed–up factor gain which can be obtained may not be worth increasing the design complexity of the simulator so much.

10.5.5 Variable Clearing

After calculating the result of the production functions of the applied programs, the next step on the algorithm is to clear the values of the variables on which these production functions depend. In practical terms, it means setting the value of these variables to 0. For doing so, each thread on the simulator has an associated production function element. In terms of implementation, each thread whose has an associated index i in the production function arrays. Thus, on every thread, the following operations are performed:

- Check its position i in *Program applications*. If the value of this position is *Inactive*, do not execute the following steps and exit the kernel. If the value of this position is *Active*, execute the following steps.
- Check if its position i in *Production function node types* is equal to *Variable*. In other case, abort the thread.
- Access its position i in *Production function variables*. Let j be the value of this coefficient.
- Set position j in *Variables* to 0.

10.5.6 Repartition Protocol Application

The last step in the algorithm consists on distributing the result of the production functions. For each thread, this implies reading the value stored in *Production function results* and distributing it over its program's contributed variables. As the normalization of coefficients is performed at the beginning of the algorithm, this step only entails multiplying this read value by the associated coefficient of each variable in the repartition protocol and adding the result of the multiplication to this variable. Before explaining in detail the implementation of this process, it is important to introduce the data structures used to represent the repartition protocols of the simulated model. Each repartition protocol is stored as a region in two arrays. Thus, each pair *coefficient–variable* has an associated index, which corresponds to an associated position in each of these arrays.

Repartition protocol coefficients: This array contains the coefficients associated to each variable existing in repartition protocols. On the repartition protocol step, the content of this array is already normalized, as it is performed at the beginning of the algorithm (see subsection 10.5.2).

Repartition protocol variables: This array contains the indexes of the variables to which the repartition protocols are contributed. These indexes are used to access the array *Variables*, in order to obtain their current value.

In terms of implementation, the distribution of the result of the production function of each program is performed the following way. Each thread has an associated pair *coefficient-variable* assigned. In terms of implementation, a position i in the repartition protocol arrays is asigned to each thread. Taking into account this consideration, each thread performs the following operations:

1. Check its position i in *Program applications*. If the value of this position is *Inactive*, do not execute the following steps and exit the kernel. If the value of this position is *Active*, execute the following steps.
2. Access its position i in *Repartition protocol coefficients*. Let c be the value of this coefficient.
3. Access the position o of the program of its repartition protocot in *Production function results*. Let f be the value of this result.
4. Perform the multiplcation of these values. Let $m = c \times f$.

5. Access its position i in *Repartition protocol variables*. Let v be the value of this position.
6. Add m to position v in *Variables*.

It is important to notice that, in this step, the theoretical speed-up factor can be greater than 1, in the case that there exist programs in the simulated model in which the number of pairs *coefficient–variable* is greater than 1. In contrast to the case of production functions, this is usual in the studied models [37], so a greater theoretical speed-up factor can be obtained in this stage.

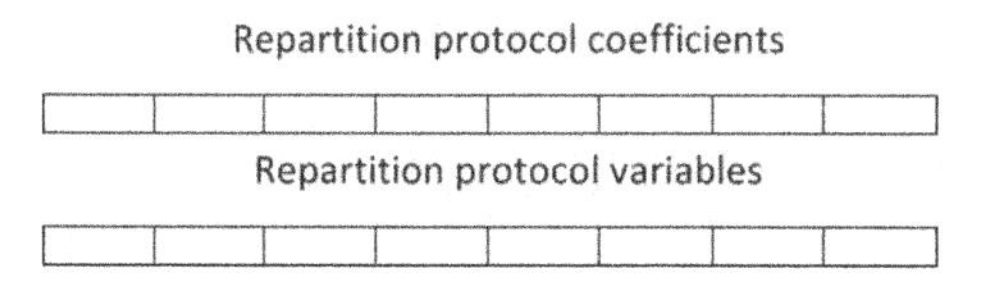

Fig. 10.7 Data structure for production functions

10.5.7 Execution of a Simulation Step

As described in the former subsections, the execution of a simulation step consists of the checking and application of programs for a predefined number of steps. This number of steps, as well as the model to simulate, are specified as inputs to the simulator. In the case that the model simulated defines a number of steps, then this number prevails over the one given as input. The simulation of a model is performed by executing the following steps:

1. Normalize the repartition coefficients, as described in subsection 10.5.2.
2. For each simulation step, perform the following operations:
    a. Assign a program to each thread. This is done by using the indexes of the threads in the *CUDA* programming model.
    b. Each thread checks if its program is to be applied, as described in subsection 10.5.3.
    c. If its program is to be executed, each thread calculates its production function, as described in subsection 10.5.4.2.
    d. If its program is to be executed, each thread clears the values of those variables which depend on the production function of the program (that is, consumes its values), as described in subsection 10.5.5.
    e. Assign a pair *coefficient–variable* from each repartition protocol to each thread. This is also done by using the indexes of the threads in the *CUDA* programming model.
    f. If its repartition protocol's program is to be executed, each thread distributes the result of the corresponding production function according to the associated pair, as described in subsection 10.5.6.

10.5.8 Remarks on the Simulator

This simulator will be published under open source license. It can be used for simulating complex distributed processes modelled with ENPSs. Therefore, several robot behaviors can be simulated in parallel (for example, a robot could avoid obstacles, follow another robot or look for a target at the same time). The synchronization at the same time between several behaviors of one robot is done by the help of the enzyme variables which can be used as stop conditions [37]. Apart from simulating several behaviors for only one robot in parallel, the simulator could be used to simulate interaction and cooperation between several robots in complex distributed robotic systems.

10.6 Simulator Performance

10.6.1 Simulator Workflow

In order to ease the simulation of ENPS models, the simulator takes an input file describing an ENPS in XML format. The XML format used is the one accepted by SNUPS [36], a previously existent sequential simulator for ENPSs. This way, the reusability of the models is improved, as the same file can be used with independence to the selected simulator, be it SNUPS and on the GPU-based one introduced, without any change in the XML file format. Hence, there is no need to change the file format, in the case that the same ENPS is to be simulated on both simulators.

Thus, in order to simulate an ENPS, one needs to encode it on the same XML format as it is required on SNUPS. Once this P system is encoded, the resulting file can be parsed by the GPU simulator. After the parsing process, the simulation is performed. Eventually, the information is displayed on the command prompt. Figure 10.6.1 shows a graphical representation of this process.

10.6.2 Performance Comparison

All parallel parts of the algorithm are executed with a degree of parallelism at least equal to the number of programs of the simulated model. The degree of parallelism can be even greater when the repartition protocol stage is applied. Hence, a theoretical acceleration of at least the number of programs of the model could be reached, if compared to the runtime of sequential simulators. In real terms, the simulator was tested by using an ENPS model of obstacle avoidance [37] as an example. These models were simulated by using SNUPS [36]. Then, the resulting runtimes were compared with the GPU simulator runtimes, in order to get an approximate speed-up. In the specific case of the obstacle avoidance model, the total number of programs is 41 [37]. Hence, an acceleration of at least 41 is theoretically expected in this case, if compared to sequential ENPSs simulators [36].

The novelty of ENPSs as a computing model [36] accounts for the need to generate *ad-hoc* case studies for the simulator. That is, it is not possible to find an

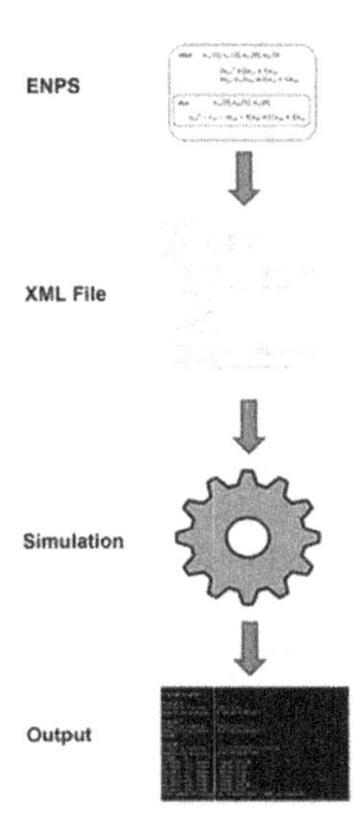

Fig. 10.8 Workflow of the simulator

extensive collection of case studies in the literature. Thus, the authors needed to generate them in order to measure the experimental performance of the GPU simulator proposed. In practice, the simulator performance has been tested by using an obstacle avoidance model [37]. Taking this model as a starting point, some case studies have been generated. All of them share the same programs, variables and membrane structure of the obstacle avoidance model proposed in [37]. Thus, these models consist of 9 membranes, 41 programs and 29 variables each [37]. The only differences between these case studies consist of the initial values of the variables associated to the membranes.

Model number	SNUPS	GPU	Acceleration
1	36.3702	6.7286	5.4053
2	14.9084	6.6304	2.2484
3	14.9040	7.7268	1.9288
4	26.3204	6.8255	3.8561
5	15.2276	6.4188	2.3723
6	18.9548	6.5659	2.8868
7	30.7377	6.7206	4.5736
8	27.0497	7.6020	3.5582
9	15.7529	6.8335	2.3052
10	30.1695	6.6364	4.5460

Fig. 10.9 Comparison of execution times for a sequential ENPSs simulator (SNUPS) and the GPU ENPSs simulator proposed

For this purpose, 10 randomly generated models were executed. Each model was executed for 100 steps. Figure 10.9 displays the execution times for these runs. This table compares the execution times for the same models run on SNUPS [36] and the GPU simulator. The execution times are given in milliseconds. For each model, the acceleration is given as the result of the division $\frac{SNUPS\ runtime}{GPU\ runtime}$.

10.7 Conclusions

In this paper, a GPU-based simulator for ENPSs. ENPSs describe a parallel computing model with applications in artificial intelligence. This simulator might be suitable for large scale models which can be applied within the field of robotics.

The massively parallel environment provided by the GPUs is suitable for ENPSs simulations. Following this line of work, it would be interesting to simulate these models by means of GPU clusters or other parallel architectures (such as FPGAs or computer clusters). These systems might be applied to model the behavior of massive robot swarms and complex sensor networks.

ENPSs can be used to model different behaviours, such as *follow the leader*, *obstacle avoidance* and *wall following* (cf. Chap. 9 of this book). The resulting simulators could be compared in terms of execution time and performance. This comparison could help experts select the most suitable simulator for the task in hand, be it *wall following*, *obstacle avoidance*, etc.

Another interesting challenge concerning the parallel simulation of ENPS models has to do about exploring the possibility of simulating several robot behaviors in parallel on GPUs. That is, simulating situations in which robots need to achieve more than one objective at the same time. These simulations could help to recreate scenarios in which robots need to perform multi-objective tasks.

Another important open problem concers the integration of the simulator into user-oriented software platforms. This integration will ease the use of the simulator by Membrane Computing experts, thus improving the human-computer interaction experience. Some examples of end-user software frameworks for simulating P systems are SNUPS[36] and P-Lingua[13].

Another important point with which to deal has to do with a more exhaustive evaluation of the performance of the simulator. Whilst the shallow performance evaluation included in this paper shows an average speed-up factor of 3x if compared to the *Java* simulator SNUPS, in order to assess the real speed–up factor to be reached by the simulator it is necessary to develop larger models and compare their runtimes not only with *Java* or other virtual machine-based programming languages, but also with languages on a lower level of abstraction, such as C or Fortran.

Nevertheless, the current models have such a small number of programs that these low-level simulators could yield better runtimes than the GPU simulator, as they are free from the overhead regarding the distribution of tasks among the GPU threads. In other words, the GPU simulator is expected to yield a better performance only when the number of programs is considerably high, that is, about thousands of programs per model. In order to asses the performance in these cases, it is necessary to extend the models currently found in the literature up to new models with thousands of programs. On these models, the GPU simulator is expected to yield lower execution times not only compared to SNUPS execution times, but also to the times obtained by C and Fortran simulators.

Acknowledgements. Manuel García-Quismondo, Mario J. Pérez-Jiménez and Luis F. Macías Ramos are supported by project TIN 2009-13192 from "Ministerio de Ciencia e Innovación" of Spain, co-financed by FEDER funds. Manuel García-Quismondo and Mario J. Pérez-Jiménez are also supported by "Proyecto de Excelencia con Investigador de Reconocida Valía P08-TIC-04200" from Junta de Andalucía. Manuel García-Quismondo is also supported by the National FPU Grant Programme from the Spanish Ministry of Education. Mario J. Pérez-Jiménez is also supported by project TIN2008-04487-E from the Spanish Ministry of Science and Education through the Complementary Action TIN2008-0448-E/TIN.

References

1. Antonelli, G., Chiaverini, S., Fusco, G.: A calibration method for odometry of mobile robots based on the least-square technique: Theory and experimental validation. IEEE Transactions on Robotics 21(5), 994–1004 (2005)
2. Cabarle, F.G.C., Adorna, H., Martínez, M.A.: A Spiking Neural P System Simulator Based on CUDA. In: Gheorghe, M., Păun, G., Rozenberg, G., Salomaa, A., Verlan, S. (eds.) CMC 2011. LNCS, vol. 7184, pp. 87–103. Springer, Heidelberg (2012)
3. Cardona, M., Colomer, M.A., Pérez-Jiménez, M.J., Sanuy, D., Margalida, A.: Modeling Ecosystems Using P Systems: The Bearded Vulture, a Case Study. In: Corne, D.W., Frisco, P., Păun, G., Rozenberg, G., Salomaa, A. (eds.) WMC9 2008. LNCS, vol. 5391, pp. 137–156. Springer, Heidelberg (2009)
4. Cecilia, J.M., García, J.M., Guerrero, G.D., Martínez-del-Amor, M.A., Pérez-Hurtado, I., Pérez-Jiménez, M.J.: Simulation of P systems with Active Membranes on CUDA. Briefings in Bioinformatics 11(3), 313–322 (2010)
5. Cecilia, J.M., García, J.M., Guerrero, G.D., Martínez-del-Amor, M.A., Pérez-Hurtado, I., Pérez-Jiménez, M.J.: Simulating a P system based efficient solution to SAT by using GPUs. The Journal of Logic and Algebraic Programming 79, 317–325 (2010)
6. Ceterchi, R., Mutyam, M., Păun, G., Subramanian, K.G.: Array-rewriting P systems. Natural Computing 2(3), 229–249 (2003)
7. Christinal, H.A., Díaz-Perni, D., Gutiérrez-Naranjo, M.A., Pérez-Jiménez, M.J.: Thresholding of 2D Images with Cell-like P Systems. Romanian Journal of Information Science and Technology (ROMJIST) 13(2), 131–140 (2010)
8. Christinal, H.A., Díaz-Pernil, D., Gutiérrez-Naranjo, M.A., Pérez-Jiménez, M.J.: Tissue-like P Systems Without Environment. In: Martínez del Amor, M.A., Păun, G., Pérez-Hurtado, I., Riscos, A. (eds.) Proceedings of the Eighth Brainstorming Week on Membrane Computing, pp. 53–64. Fenix Editora (2010)
9. Colomer, M.A., Lavín, S., Marco, I., Margalida, A., Pérez-Hurtado, I., Pérez-Jiménez, M.J., Sanuy, D., Serrano, E., Valencia-Cabrera, L.: Modeling Population Growth of Pyrenean Chamois (Rupicapra p. pyrenaica) by Using P-Systems. In: Gheorghe, M., Hinze, T., Păun, G., Rozenberg, G., Salomaa, A. (eds.) CMC 2010. LNCS, vol. 6501, pp. 144–159. Springer, Heidelberg (2010)
10. Conforth, M., Meng, Y.: An Artificial Neural Network Based Learning Method for Mobile Robot Localization. Robotics Automation and Control 6 (2008)

11. Dong, J., Liu, B., Peng, K., Yin, Y.: Robot Obstacle Avoidance based on an Improved Ant Colony Algorithm. In: Zhou, S.M., Wang, W. (eds.) Proceedings of WRI Global Congress on Intelligent Systems (GCIS 2009), pp. 103–106. IEEE Computer Society (2009)
12. Du, R., Zhang, X., Chen, C., Guan, X.: Path Planning with Obstacle Avoidance in PEGs: Ant Colony Optimization Method. In: Zhu, P. (ed.) International Conference on Cyber, Physical and Social Computing (CPSCom), pp. 768–773. IEEE Computer Society (2010)
13. García-Quismondo, M., Gutiérrez-Escudero, R., Martínez-del-Amor, M.A., Orejuela-Pinedo, E., Pérez-Hurtado, I.: P-Lingua 2.0: A software framework for cell-like P systems. International Journal of Computers, Communications and Control 4(3), 234–243 (2009)
14. Garland, M., Le Grand, S., Nickolls, J., Anderson, J., Hardwick, J., Morton, S., Phillips, E., Zhang, Y., Volkov, V.: Parallel computing experiences with CUDA. IEEE Micro 28(4), 13–27 (2008)
15. Gill, M.A.C., Zomaya, A.Y.: Genetic algorithms for robot control. In: IEEE International Conference on Evolutionary Computation, p. 462. IEEE Computer Society (1996)
16. Gerstner, W., Kistler, W.: Spiking Neuron Models. Single Neurons, Populations, Plasticity. Cambridge Univ. Press (2002)
17. Ibarra, O., Pérez-Jimenez, M.J., Yokomori, T.: On spiking neural P systems. Natural Computing 9(2), 475–491 (2010)
18. Ionescu, M., Paun, G., Yokomori, T.: Spiking neural P systems. Fundamenta Informaticae 71 (2-3), 279–308 (2006)
19. Ionescu, M., Paun, G., Pérez–Jiménez, M.J., Rodríguez-Patón, A.: Spiking Neural P systems with several types of spikes. International Journal of Computers, Communications & Control 4(4), 648–656 (2011)
20. Ivankjo, E., Komšić, I., Petrović, I.: Simple Off-Line Odometry Calibration of Differential Drive Mobile Robots. In: Proceedings of 16th International Workshop on Robotics in Alpe-Adria-Danube Region - RAAD, pp. 164–169 (2007)
21. Khatib, O.: Real-time obstacle avoidance for manipulators and mobile robots. The International Journal of Robotics Research 5(1), 90–98 (1986)
22. Maass, W.: Computing with spikes. Special Issue on Foundations of Information Processing of TELEMATIK 8(1), 32–36 (2002)
23. Maass, W., Bishop, C. (eds.): Pulsed Neural Networks. MIT Press, Cambridge (1999)
24. Nickolls, J., Buck, I., Garland, M., Skadron, K.: Scalable parallel programming with CUDA. Queue 6(2), 40–53 (2008)
25. NVIDIA. NVIDIA CUDA Programming Guide 2.0 (2008)
26. Pan, L., Pérez-Jiménez, M.J.: Computational complexity of tissue-like P systems. Journal of Complexity 26(3), 296–315 (2010)
27. Pan, L., Paun, G., Pérez-Jiménez, M.J.: Spiking neural P systems with neuron division and budding. Science China. Information Sciences. 54(8), 1596–1607 (2011)
28. Pan, L., Wang, J., Hoogeboom, H.J.: Asynchronous Extended Spiking Neural P Systems with Astrocytes. In: Gheorghe, M., Păun, G., Rozenberg, G., Salomaa, A., Verlan, S. (eds.) CMC 2011. LNCS, vol. 7184, pp. 243–256. Springer, Heidelberg (2012)
29. Paun, G., Paun, R.: Membrane Computing and Economics: Numerical P Systems. Fundamenta Informaticae 73(1-2), 213–227 (2006)
30. Paun, G.: Computing with membranes. Journal of Computer and System Sciences 61(1), 108–143 (2000)
31. Paun, G.: Membrane Computing. An Introduction. Springer, Heidelberg (2002)

32. Paun, G.: Computing with Membranes. Turku Centre for Computer Science, Turku, Finland, vol. 208 (1998)
33. Paun, G., Pérez-Jiménez, M.J., Riscos, A.: Tissue P systems with cell division. International Journal of Computers, Communications & Control 3(3), 295–303 (2008)
34. Paun, G., Paun, R.: Membrane computing as a framework for modeling economic processes. In: Proceedings of the Seventh International Symposium on Symbolic and Numeric Algorithms for Scientific Computing (SYNASC 2005). IEEE Computer Society, Washington DC (2006)
35. Păun, A., Păun, G.: The power of communication: P systems with symport/antiport. New Generation Computing 20(3), 295–305 (2002)
36. Pavel, A., Arsene, O., Buiu, C.: Enzymatic Numerical P Systems - A New Class of Membrane Computing Systems. In: Proceedings 2010 IEEE Fifth International Conference on Bio-inspired Computing: Theories and Applications (BIC-TA 2010), pp. 1331–1336. IEEE Computer Society, Liverpool (2010)
37. Pavel, A., Buiu, C.: Using enzymatic numerical P systems for modeling mobile robot controllers. Natural Computing (2011) (in press)
38. Romero, F.J., Pérez-Jiménez, M.J.: A model of the Quorum Sensing System in Vibrio Fischeri using P systems. Artificial Life 14(1), 95–109 (2008)
39. Takizawa, H., Koyama, K., Sato, K., Komatsu, K., Kobayashi, H.: CheCL: Transparent Checkpointing and Process Migration of OpenCL Applications. In: International Parallel and Distributed Processing Symposium (IPDPS 2011), pp. 864–876. IEEE Computer Society, Anchorage (2011)
40. `http://www.gpgpu.org`
41. `http://www.nvidia.com/object/cuda_home_new.html`
42. ISI web page, `http://esi-topics.com/erf/october2003.html`

Chapter 11
How to Design an Autonomous Creature Based on Original Artificial Life Approaches

Pavel Nahodil and Jaroslav Vítků

Abstract. We introduce new approaches for creating of autonomous agents. The life of such creatures is very similar to the animal's life in the Nature, which learns autonomously from the simple tasks towards the more complex ones and is inspired by AI, Biology and Ethology. We present our established design of artificial creature, capable of learning from its experience in order to fulfill more complex tasks, which is based mainly on ethology. It integrates several types of action-selection mechanisms and learning into one system. The main advantages of the architecture is its autonomy, the ability to gain all information from the environment and decomposition of the decision space into the hierarchy of abstract actions, which dramatically reduces the total size of decision space. The agent learns how to exploit the environment continuously, where the learning of new abilities is driven by his physiology, autonomously created intentions, planner and neural network.

11.1 Introduction — History of Artificial Life Research

Man's strive to copy the creations of Mother Nature reaches deep into the history. Originally, the power to create artificial copies of creatures and humans was exclusive to gods. In Greek mythology Hephaestus, the god of fire and patron of all craftsmen, used to create mechanical servants, ranging from intelligent, golden handmaidens to more utilitarian three-legged tables that could move upon their free will. The stories about the artificial beings occur in many different cultures across the continents, for example in China or Egypt. With the advances in science like mathematics and physics the power to create an artificial being shifted from gods to humans. In the wake of our era, around 0 AD - first designs of mechanical creatures were proposed. One of those was mechanical bird called "The Pigeon", powered by

Pavel Nahodil · Jaroslav Vítků
CTU in Prague, FEE, Department of Cybernetics, Technická 2, 166 27 Prague 6, Czech Republic
e-mail: {pavel.nahodil,vitkujar}@fel.cvut.cz

J. Kelemen et al. (Eds.): Beyond Artificial Intelligence, TIEI 4, pp. 161–180.
DOI: 10.1007/978-3-642-34422-0_11 © Springer-Verlag Berlin Heidelberg 2013

steam. It was postulated by Archytas of Tarentum. In the middle ages, the search for a mechanical copy of life continued in form of science and magic. One of the greatest scientist of that age, Leonardo da Vinci, proposed a mechanical knight who able to sit up, wave its arms and move its head and jaw. On the other hand, utilizing magic describes the legend of Rabbi Juddah's creation of Golem, an artificial being made out of the clay. In the beginning of 20th century the invention of semiconductors gave the design of mechanical beings whole new dimension. The word "robot" was first used for the artificial being in Josef Čapeks play called *Rossum's Universal Robots* (R.U.R.) [1],[2]. In the second half of the 20th century robotics as a science branch started to advance rapidly. Robots became smaller, smarter and autonomous. Scientist all over the world began to focus on various areas of robot control starting with the space orientation, navigation through recognition, adaptation, up to the behavior and team cooperation [3].

The autonomous functioning of a robot in real environment meets many challenges. The control architecture must give the robot ability to react timely with respect to the local disturbances and uncertainties, while adapting to more persistent changes in environmental conditions and task requirements [4]. The inherent problem in this area of research is that considerable work effort is required to equip robots with adequate means for sensing (sensors) and actuation (effectors). Recognition and transformation of data in noisy and voluminous environment poses an obstacle in the robot design. Thus, to study control architecture the research moved from real environments to virtual ones. The term "agent" replaced the term "robot". The research of behavior no longer needs a physical robot, the virtual representation of the robot can provide the same level of embodism as s real one. For these virtual robots in analogy with the *multi-agent systems* (MAS), the term "agent" stated to be used [5]. These two fundamentally different approaches merge by the selection of common name "agent". MAS originally used the top-down approach, focused on planning, problem solving which we can consider as a high level function of some animals and also humans. On the contrary the bottom-up approach used in robotics and also by nature in the simple organisms is focused on reactions to the stimuli. This approach uses emergence as a tool for creating more complex and complicated behavior by chaining the most basic reactions together. By joining these two approaches together with a meaningful tradeoff between theirs pros and cons proved to be a very interesting approach. On the top of this, in last decade turned to the ability to predict future changes and preparation for them. The term anticipation begun to be used in this topic.

The research presented here builds upon knowledge from several scientific fields. Ethology / Biology contributed by the various examples and experiments with animals. The studies of interest focused on the explanation of the behavior and the mechanism that produces and selects the behavior not only as a reaction to the immediate state but also with regard to the estimated future states and preparation for them. Control engineering provided a framework for formalizing, describing, controlling and modeling systems them and also for estimating their future states. Classical Artificial Intelligence offered the high deliberative functions like learning, planning and reasoning among others.

The goal, to create an autonomous robotic system capable of flawless operation in the real environment full of disturbances and unpredicted events, have not been reached. The complexity and observability of the real environment is simply too high to be exactly described by anything less complex than the environment itself. Since the given problem cannot be precisely described by the mathematics, other approaches which take uncertainty and incomplete models into account must be used. With regard of above mentioned goal we propose the best chance of fulfilling it is in constructing something relatively simple, yet capable of autonomous learning and using this new knowledge to improve itself. There is a working group at the Department of Cybernetics at the CTU in Prague, which have been trying to find and describe the possible methods for solving this problem since nineties. Both authors, members of this group, investigated the possible architectures of autonomous agents partially capable of learning and self improvement. This paper presents the latest results in the research domain, called *Artificial Life* (ALife).

11.2 State-of-the-Art — Comparison to Similar Architectures

This text will commence with a brief description selected architectures which use similar ideas for control and/or learning to our architecture. Our architecture differs by an incorporation of Biological and Ethological principles to the design. The other significant differences will be discussed in the following sections as well.

11.2.1 Integrating Neural Networks and Knowledge-Based Systems for Intelligent Robotic Control

The first architecture is used to control robotic hand (manipulator) and tries to integrate deliberative and reflexive control. This intelligent control system uses *Cerebellar Model Articulation Controller* (CMAC) [6], a Biology-inspired learning algorithm for robotic control using the *Artificial Neural Networks* (ANNs). The CMAC uses similar principles with the reinforcement learning, where the learning is supervised by the knowledge-based subsystem in this case.

The scheme of entire system is in the Fig.11.1, the agent controlled by this system learns how to use the manipulator with two joints. The course of operation of this system is as follows: first the knowledge-based system components determine how to solve a given control objective using rules and algorithms within its knowledge base. Then teach the neural network is taught how to accomplish tasks by using observation and generalization of knowledge-based task execution. The knowledge-based system components continuously evaluate neural network performance and reengage rule-based control whenever errors occur due to changes in the dynamic system. The knowledge-based subsystems thereby ensure proper task completion while relearning takes place within the neural network [7].

The use of neural network provides the system with ability to predict and generalize, while the rule-based subsystem is able to create more complicated plans and teach the network to execute them. This means that architecture is able to make a

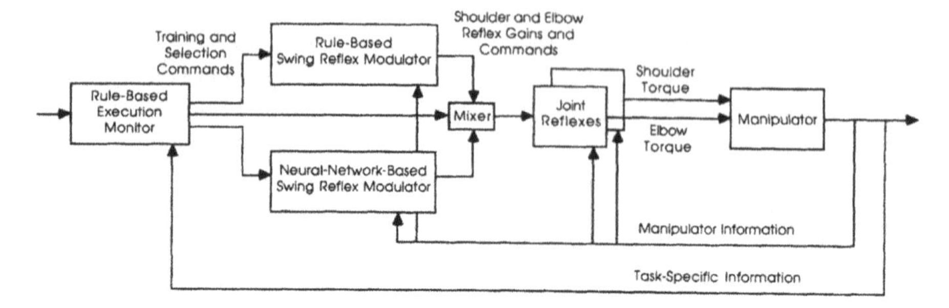

Fig. 11.1 Block diagram of possible approach of combination knowledge-based and neural network-based systems for controlling of the robotic hand. Taken from [7].

binary decision whether to use knowledge-based or network based decision making. Similarly to this system, our architecture uses supervised learning of ANN by the different subsystem. In contrast to the RL-TOPs, our architecture is more universal and does not suffer from the big gap between classical and connectionist approach. We use hierarchy of *Reinforcement Learning* (RL) actions. This hierarchy of actions (from abstract towards those primitive ones) provides a fluent transition between deliberative planning and purely reactive control generated by the neural network. By means of this principles, the gap between two different control mechanisms becomes blurred.

11.2.2 Reinforcement-Learning Teleo-Operators

In our architecture we use the combination of Reinforcement Learning and planning subsystems. A approach based on the analogous idea, called *Reinforcement-Learning Teleo-Operators* (RL-TOPs), already exists. It was first presented in [8]. The main objective of this approach is to speed up learning process by decomposition of complex tasks into hierarchy of simpler behaviors, which can be reused later. It is based on *Teleo-Reactive planning* system (TR) presented in [9], in which the planning system works with a tree of nodes. Each node corresponds to one particular state, while the root node represents the goal. Connections between nodes represent actions. The second main principle this approach uses is the *Teleo-Operator* (TOP), which means a durative action a described by its pre-image π_a and its effects λ_a, the TOP then denotes transformation:

$$a : \pi_a \rightarrow \lambda_a \tag{11.1}$$

which means that if the action is executed while the π_a is fulfilled the λ_a can eventually become true. Before the simulation, the user has to provide the following a priori knowledge:

Low-level state representation based on the robot sensors
Actions - a set of primitive low-level actions

High-level state description - language that will be used by the planner. Example of high-level state description is in the Fig.11.2.
Goal description - conditions that are fulfilled in the goal state
RL-TOPs library - a set of behavior descriptions in the form of $\langle \pi_a, a, \lambda_a \rangle$. Example of this library is in the Fig.11.3.

The agent learns the given set of these low-level behaviors using the multiple RL engines, and each behavior is represented as an action on the high level. The example of the use is described in the Fig.11.2 and Fig.11.3, where the agent has to reach the room no.4. The planner is used for two main purposes here, the first one is the creation of the plan (that is: building of the action hierarchy). The other one is used for learning during the plan execution itself; the planner specifies whether to reinforce given behavior or not. The reinforcement is generated in case that the post-effecs λ_a have been fulfilled [10].

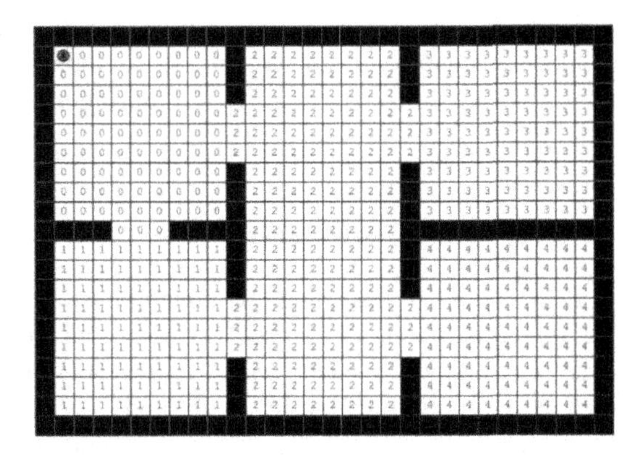

Fig. 11.2 Example use of RL-TOPs architecture, predefined high-level state knowledge which assigns a ID to each (state in the) room. Taken from [8].

This system uses a planner for some high-level decision making and also for teaching the RL engines, where each engine corresponds to some low-level behavior [8]. Our architecture also uses a combination of RL and a planner, but in this current version the planner is not used for behavior reinforcement (teaching). Compared to RL-TOPs, the main advantage of our architecture is that we do not need practically any knowledge a priory. This means that the equivalents of RL-TOPs library and the high-level state description is obtained autonomously during the agent's existence, based solely on the concrete domain of the use.

11.3 Architecture with Elements of Hierarchy, Abstraction, Reinforcements and Motivations

David Kadleček, in his dissertation thesis [11], presented the system called *"Hierarchy, Abstraction, Reinforcements, Motivations Agent Architecture"* (HARM),

RL-TOP	pre-image	post-condition
go02	room(0)	room(2)
go20	room(2)	room(0)
go12	room(1)	room(2)
go21	room(2)	room(1)
go32	room(3)	room(2)
go23	room(2)	room(3)
go42	room(4)	room(2)
go24	room(2)	room(4)

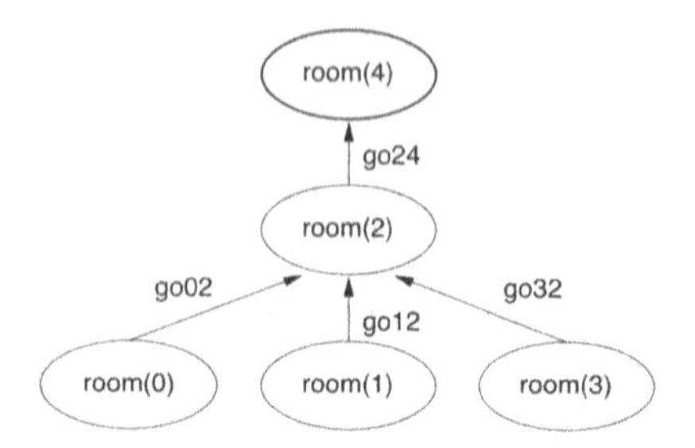

Fig. 11.3 Example use of RL-TOPs architecture, the agent's goal is to reach the given room. Knowledge contained in the RL-TOPs library (on the left) describing the set of available behaviors (actions). On the right there is a tree of high-level actions created by the planner in order to reach the goal state described by the predicate *room(4)*. Taken from [8].

which was inspired mainly in Ethology. This system uses for action selection Reinforcement Learning mechanism and it is capable of autonomous creation of action hierarchy based on the various reinforcements received during the interaction with the environment [12], [13], [14].

11.3.1 Main Ideas of the HARM Approach

According to the HARM, the agent has some predefined physiological state space modeled by the *Multiple Input-Multiple Output* (MIMO) dynamic system. This space contains several variables, where each variable represents some of the agent's needs. This space contains two important areas: a limbo and a purgatory. Limbo area represents optimal conditions: if the agent is in this area no motivation is produced. On the other hand, the purgatory area represents critical conditions, in this area the motivation for the corresponding behavior increases exponentially. If the agent manages to actively move the physiological state towards the limbo area, the appropriate reinforcement is generated. In response to this setup, the agent creates connections between his behavior and physiology and learns how one influences the other. The agent is able to learn how to select appropriate actions in order to continuously optimize his internal conditions, the principle is called in biology homeostasis. Fig.11.4 shows an example of physiological state space with two basic needs, denoted as x_1 and x_2. This, for example, can be used (in case of a robotic system) as a need to recharge the battery or to maintain the safe temperature.

The agent learns (using Q-learning algorithm) new behaviors in order to maintain his physiology in optimal conditions. If the previously unknown reinforcement is received, new decision space is partitioned from the original decision space. This new decision space is connected to the corresponding source of motivation and thus can be motivated by its own physiological variable. As a result, the agent uses multiple return predictors where each corresponds to some particular behavior. The scheme of this approach is depicted in the Fig.11.5 where the original control is decomposed into the hierarchy of small decision spaces.

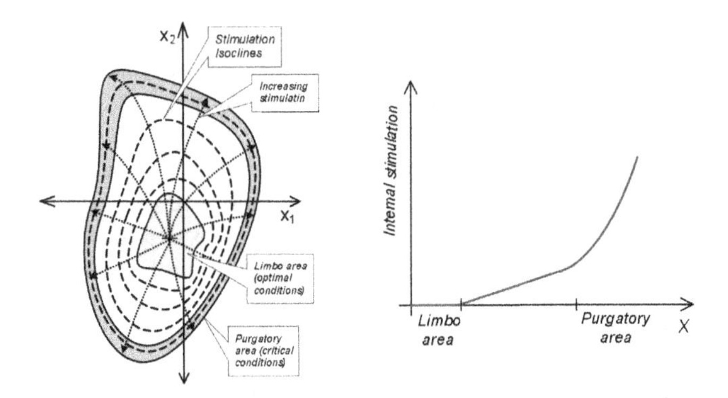

Fig. 11.4 Physiological state space containing several variables (x_1 and x_2 here) which represent some agent's needs. This dynamical system serves as a source of motivation; the agent learns how to optimize these objective functions.

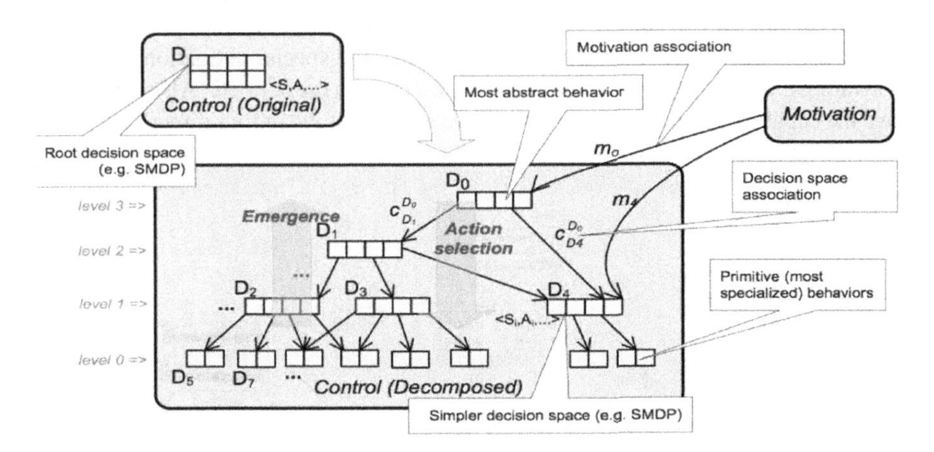

Fig. 11.5 Hierarchically decomposed decision space by the HARM system. The root node corresponds to the most abstract behaviors, whereas the leaf nodes correspond to the lowest behaviors-primitive actions.

One of the important advantages of this hierarchical reinforcement learning system is the fact that each decision space contains only a subset of actions and environmental variables, which dramatically reduce the total size of search space used by the return predictors (RL engines). These sets of variables and actions (which should be contained in the decision spaces) are maintained by using four main principles: sub-spacing, variable promoting, behavior associating and variable removing, for more details please refer to [11]. Also, the Q-Learning algorithm used in each decision space provides online trade-off between the exploration and knowledge exploitation. Another interesting fact is that during the action selection phase, the return predictors operating in hierarchy can either cooperate or compete. Cooperation can be observed in a case of two similar goals, or in a case that one abstract behavior is composed of some less abstract actions. In this case, the motivation for the currently selected action is passed lower into the hierarchy and added to the selected decision space. The competition between return predictors can be observed in case of two antagonistic strategies, like: in the case when energy source is located near the perceived danger zone, the two competing motivations are: need to supply the energy and escape. The motivation to execute more and more primitive actions is passed towards the bottom of the hierarchy. This means that the final selection of concrete primitive action is composed as a sum of intentions produced by all parent nodes in the behavior hierarchy (see Fig.11.5).

11.3.2 Sample Experiment: The Treasure Problem

In this selected experiment, the agent's task is to get to the treasure locked behind the door. In order to open the door the agent has to put the stones onto the buttons in a specific order. Beside learning the task, the agent needs to drink and eat in order to survive. Before the commencement of a simulation, the agent's physiological state space is equipped with variables: water, food and special obligation variables motivating the agent to pick and drop the stones, to open the door etc. Also, the agent have the capability of the following primitive actions: move in four directions, pick, drop, eat and drink.

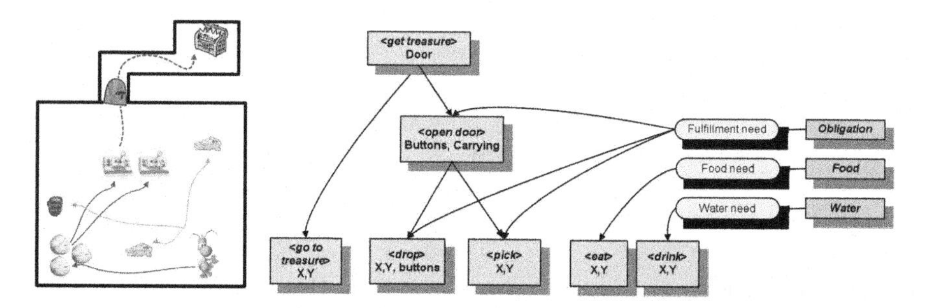

Fig. 11.6 The treasure problem: agent has to reach the treasure. In order to open the door he must put the stones onto the switches in the specified order. Agent has to follow two physiological needs: hunger and thirst. On the right side there is depicted the result of creating the action hierarchy, where in each decision space is name of the behavior and a list of variables that are considered by its return predictor.

On the left in the Fig.11.6 we can see the problem description, on the right side there is depicted the resulting action hierarchy with connections of decision spaces to its own sources of motivations. We can see that the reaching the treasure can be decomposed to more primitive behaviors "open the door" and "go to treasure". The action "open the door" can be further decomposed into two subtasks "pick stone" and "drop stone". This autonomously created action hierarchy efficiently represents the problem structure and the agent is able to reach the treasure and simultaneously eat or drink if necessary.

11.4 Proposed Autonomous Creature

The latest result of our research is called "*An Artificial Creature Capable of Learning from Experience in Order to Fulfill More Complex Tasks*" [10] and it is an autonomous system composed of several different approaches gained mainly from the fields of Artificial Intelligence, ALife and Ethology. When designing the agent, the main goals were: to reach the maximum degree of autonomy, the ability to deal with a complex and dynamically changing environments (similar to the real world), and a maximum domain independency. The goal was to create an autonomous agent which, placed into an unknown environment, will be able to learn everything by itself, to be able to learn how to survive, to gain new information in order to live (behave) more efficiently, and to learn and explore new objects in its environment similarly to a newly born animal.

11.4.1 Brief Description of Designed Architecture

The main feature of this architecture is in its total autonomy, an ability to gain all of the information from the surrounding environment and effective information filtering and classification. Agent can operate using solely the sensory inputs and by its actuator system, thus the resulting architecture is almost fully independent on the concrete area and form of use. Consequently it is irrelevant whether the agent is embodied in some robotic system, intelligent house, or operates in some virtual environment. Thanks to the fact that all the designer has to specify is the sensory layer, actuator layer and agent's needs; this architecture is convenient especially in unknown environments, where the user needs the execution of a complex task with no a priori knowledge about it.

Agent architecture is inspired by the layered model, combining various approaches on different levels. Learning occur on-line during the agent's life along with the action selection. As a quantity and quality of learned knowledge increases, the agent exhibits consecutively more systematic behavior. The life of agent begins similarly to a newly born animal which explores new and unknown environment, learns from experiences, links the newly learned abilities to assist fulfilling its needs in order to survive and increase effectiveness of its behavior. New knowledge is learned simultaneously on various levels of abstraction using different learning approaches. These will be described in more details below.

One of the most important features is an alternative implementation of system similar to reactive and hierarchical planning. The system combines hierarchical reinforcement learning and planning engine into a domain independent hierarchical planner.

This agent could be used in the real or virtual environments where the designer has no a priori knowledge about its fundamental regularities and is able to specify only the agent's needs and several high-level goals. The agent autonomously explores the environment and attempts to "understand" the principles from simple to the complex ones in order to gain the ability to survive and fulfill the assigned goals.

According to our belief, the single kind of problem representation or approach to solving the problem is almost never sufficient. Each action made by living animals is a consequence of superposition of many different motivations, needs, emotions, intentions etc. We simulate this by connecting several of the decision and control blocks. Each of these blocks consists of one of the well-known and widely used systems, such as reinforcement learning or a planner. So instead of attempting to implement all of the decision making and learning by e.g. neural networks and expecting the emergence of high-level behaviors, we connect the different blocks in such a way that the resulting system suppresses the weaknesses of particular subsystems and exploits benefits of each more efficiently.

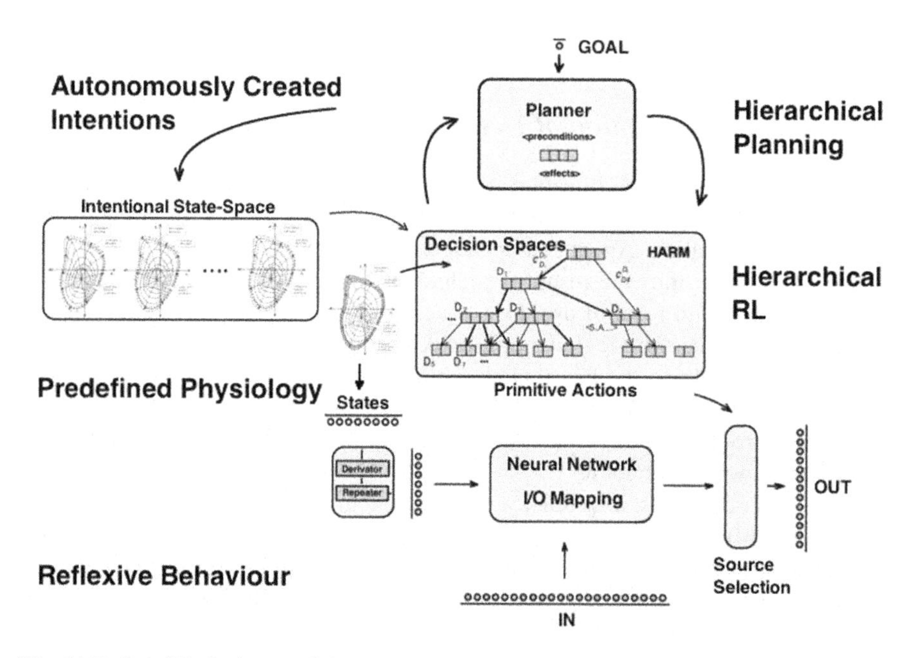

Fig. 11.7 Simplified scheme of the agent architecture

In the Fig.11.7 we can see simplified scheme of entire agent architecture Main subsystems used here are (listed from the bottom): an artificial neural network, a hierarchical reinforcement learning and a hierarchical planning. The description of their purpose in more details is below.

11.4.2 Augmenting Original Hierarchical Learning System

The core of entire architecture is HARM system presented in the previous section. Elaborating on HARM system, this is enhanced in several ways. The main improvement is a new ability of on-line learning and creation of hierarchy. The other main improvement presents the agent with a higher degree of freedom, and an ability to discover new knowledge autonomously. The course of learning will be described below.

At the beginning of the simulation, the hierarchy of abstract actions is empty. The agent acts randomly and observes whether something interesting occured. There are two main possibilities: the agent actively changes some of his physiological state variables or manages to actively change some environment variable. In the second case, the new intentional variable is created and added to the agent's intentional state space. In both cases the central HARM system executes the hierarchy update as described in the section above.

11.4.3 Intentional State Space

The intentional state space has the similar purpose as the physiological state space in the HARM, however, here the main agent's intentions are created autonomously during his life. If the agent discovers is able to actively change some environment variable (e.g. turn on the lights), the new intentional variable is created. This variable motivates the agent to learn and "train" this newly discovered behavior, which does not necessarily influence agent's physiology. Intentional state variables have its own predefined dynamics and do not include the purgatory area; this means that the agent learns these behaviors only in non-critical situations. Due to this inclusion, the agent can autonomously discover new potentialities of the environment and learn how to exploit them and optionally use this new knowledge can be reused later. This approach corresponds to learning of young animal by play.

11.4.4 Deliberative Action Selection — Planning

The other important subsystem of the architecture is the hierarchical planning engine. It is composed of classical "flat" planner operating over the hierarchy of decision spaces. One of the most important features of this architecture is the ability to represent the abstract action from the RL action hierarchy as a set of primitive actions in the *Stanford Research Institute Problem Solver* (STRIPS) language [15]. This gives the agent ability to deliberatively "think" about the actions previously learned during the interaction with the environment and to fulfill complex tasks. The advantages of hierarchical decomposition of plans are already well known, and referred for example in [16] as *Abstraction Space Hierarchy from STRIPS* (AB-STRIPS) or *Hierarchical-Task Network* (HTN) [17]. The planner can put into the effect abstract actions on selected arbitrary level(s) of action abstraction, which provide configurable plan granularity and precision. This principle is demonstrated in the Fig.11.8.

In the Fig.11.9 is depicted the principle of representing the RL abstract action (decision space) as a set of primitive actions in the STRIPS language. The goal is to look at the action *"from the outside"* and identify which informations are important. The goal of particular abstract action in the augmented HARM system is to change

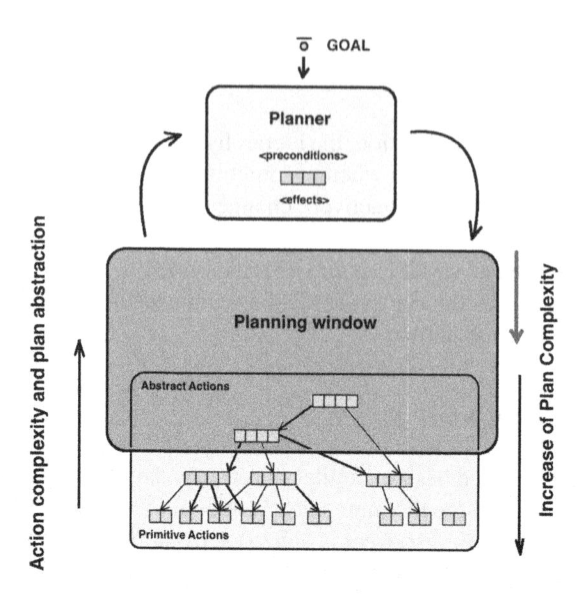

Fig. 11.8 Principle of planning over the hierarchy of decision spaces on some selected level(s) of abstraction. The abstract actions (and their "main" environment variables) are considered when creating the plan; the rest of the problem is solved on-line by the reinforcement learning when necessary.

the value of one environment variable (in case of actions motivated by intentional state space). Therefore only this "main" variable of decision space is visible to the planner. Then we can generate a set of primitive actions that cause the change of this variable (e.g. turn on and turn off the lights).

The main benefit of this solution is in the fact that the complicated outer world is pre-processed by the RL-action hierarchy, so this hierarchical planner can operate in very complex domains, while still maintaining the domain independence. This is a significant advantage against the well-known hierarchical planners. More precisely, our planner could be referred as domain self-configurable. The HARM action hierarchy serves as some "interface" between planning engine and the outer world. This interface is created autonomously and adapts to the given problem online during the agent's life.

11.4.5 Artificial Neural Network — Learning the Reflexive Behavior

The last part of architecture serves for learning of reflexive behavior and gives the agent ability to react in selected situations with necessary speed of response. This subsystem is implemented by *Artificial Neural Network* (ANN) and detects and memorizes the patterns [*situation-action*]. The ANN learns appropriate behaviors from the HARM system. When the situation is considered critical, the HARM

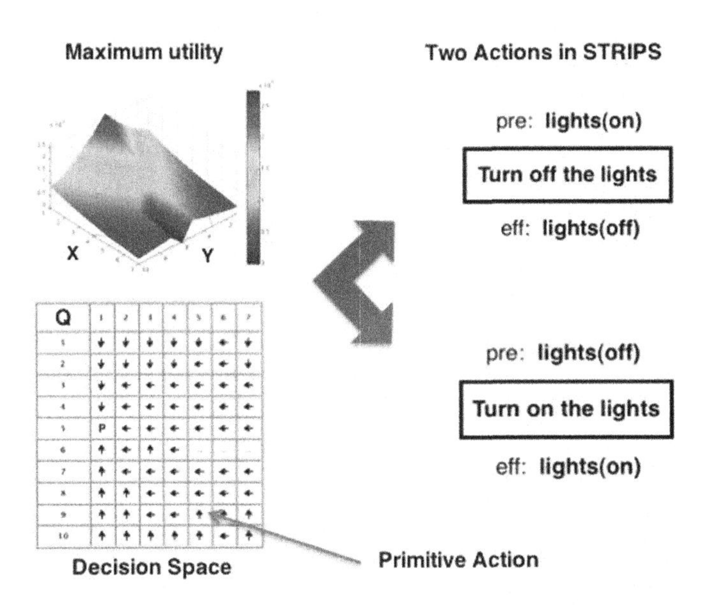

Fig. 11.9 Example of representing the decision space (abstract RL action) as two primitive actions in the format of STRIPS language. The environment variable that caused the reinforcement is the "main" one in the decision space (lights in this case) and only this variable is visible to the planning engine. Based on the possible states of this variable, a set of possible primitive actions in the STRIPS language can be generated.

system generates adequate action while ANN learns this one pattern [*actual situation -generated action*]. After some time, when the ANN exhibits small enough learning error, the agent can act reflexively. From this moment on the ANN can take the control over the agent, and generate a primitive action in some critical situations. This system provides quick reaction speed where it is important. Also this approach increases the precision of action selection; this is caused by the ability of the neural networks to generalize.

11.5 Selected Experiments

We have conducted numerous experiments in order to test the implementation of our architecture, and to compare with other existing approaches. The main focus was on testing the ability to effectively reduce the size of decision space, and to act in dynamic environments (e.g. predator-prey simulations). Lastly the focus was on the ability to create and execute plans based on the knowledge gained autonomously. We will describe conclusions from two of the experiments.

11.5.1 *Two Attractors in Complex Environment — Hierarchy Creation*

In the Fig.11.10 on the left we can see the map of the selected experimental environment which includes our agent (in the corner), five slow, randomly moving, white cows. Orange stands for the source of food, blue for a source of water. In this experiment our agent has only one goal - to survive. Based on the predefined physiology, the agent has to learn how to eat and drink, meaning that the agent has to learn two behaviors: drinking and eating, motivated from the physiological variables food and water. While the food and water source positions are static, the resulting decision spaces should contain only the agent's position, while ignoring the other (for the particular task) unimportant variables. The unimportant variables are $\langle X, Y \rangle$ positions of all cows in this case.

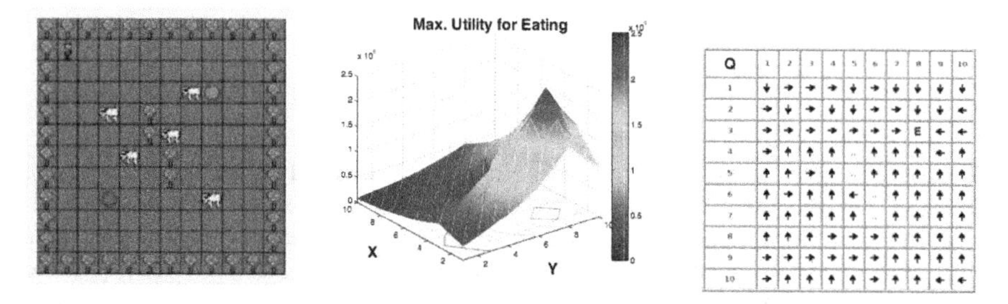

Fig. 11.10 Experiment testing the reduction of decision space size. The map (left) contains food (orange object) and water (blue object) sources and the size of the decision space was increased by other moving agents (white cows). The goal is to learn how to survive - how to maintain the water and food levels in agent's body within bounds. During his life, the agent was able to successfully decompose his behavior into two abstract actions, "eat" and "drink". For the "eating" behavior, we can see the graph of maximum utility (and the table describing the best action) corresponding to the agent's actual $\langle X, Y \rangle$ position. Matrix on the right side represents learned primitive actions (ones with the highest utility) based on the agents position. The matrix contains only 100 states, meaning a dramatic reduction of decision state space size compared to the flat RL approach

After some time, the agent was able to identify both behaviors fundamental to his survival. By using the four hierarchy creation strategies, he was able to consecutively remove all of the unnecessary variables (cow positions) from the decision spaces.

While the agent's internal variables have 4 states both, sources of food and water have 2 states. Each moving object (cow or our agent) can occupy 10×10 positions on the map, meaning that the total size of original decision space was approximately 64×10^{12} states. The agent successfully learned that the positions of particular cows can be ignored, and thus was able to reduce the number of states to 2×100 states, 100 states in each decision space corresponding to one behavior. From the Fig.11.11

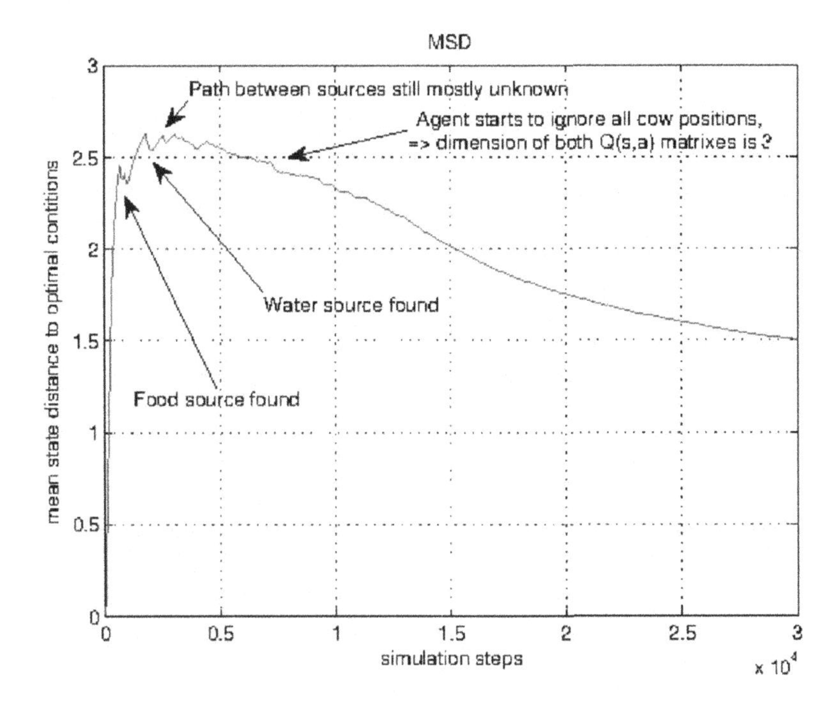

Fig. 11.11 Course of the agent's learning during the experiment. The graph represents *Mean State Distance to optimal conditions* (MSD) in time. The moment when the agent was able to find the food source, water source is visible. Later, the path between two attractors was learned. From the graph is apparent how the successful removing of the unnecessary variables from the hierarchy speeds up the learning process.

we can see how the successful removing of these unnecessary variables can speed up the learning convergence towards the stable behavior.

11.5.2 Part 1 — Creation of Hierarchy Based on Intentions

In this experiment we have tested the agent's ability to reuse autonomously gained knowledge, meaning the ability to plan and solve some complicated task based only on the knowledge learned from the interaction with the environment.

In this experiment, the user needs to pass the hallway in order to reach the goal position. The requirements are that all doors on the path are opened and the lights are turned on. The only user's a priori knowledge is that these systems can be controlled only from unknown and unreachable part of the map. So our agent is sent to find out how these systems can be controlled. The agent learns how to survive simultaneously, so after some time, the agent is physically there and able to fulfill relatively complex tasks, as, for example, "*enable passing through the hallway*", which is composed of subtasks: *"turn on the lights"*, *"open the door1"* and *"open the door2"*. The initial and the desired state of the map are depicted in the Fig.11.12 on the left. The user does not have any prior knowledge about the problem, his only knowledge is that the controls to these properties are somewhere in the unknown sector of the map. Our agent is sent to autonomously learn these principles, while he has only two physiological needs predefined and a set of the following primitive actions: move in four directions, eat, drink, press.

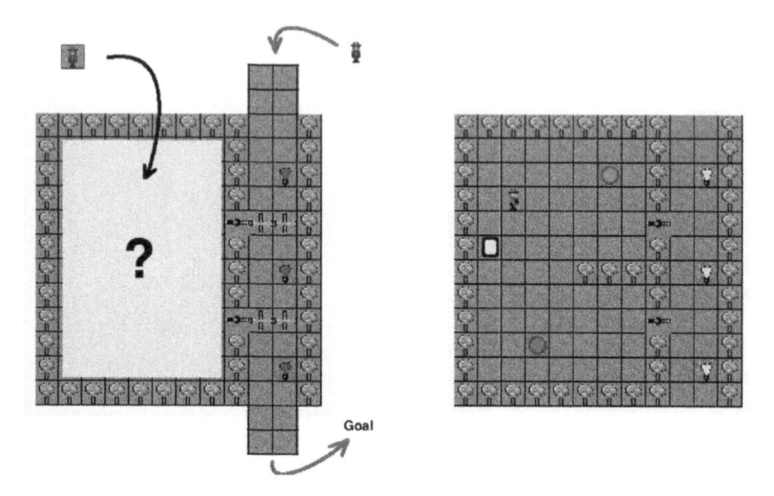

Fig. 11.12 Description of the task. User wants to be able to travel between his position and goal position through the hallway. Successful travel requires turned on lights, and all doors on the path opened. The red agent is sent to learn how to survive in this unknown environment and to discover the main principles needed to execute the task. After some time, the red agent is able not only to survive in this unknown environment, but also to fulfill the user's demands. In the right picture we can see map in desired state, also there are visible the water and food sources, silver switch controlling lights and two buttons controlling the doors are depicted on the picture as well.

After some time into the simulation, the agent was able to successfully identify and learn all five possibilities how to interact with the environment. The Fig.11.13 shows the resulting action hierarchy. It is apparent that during his existence, the agent autonomously creates three intentions motivating him to learn how to control lights and both doors. The agent is now able to successfully drink, eat, switch the lights and to open/close both doors. In the Fig.11.14 the learned primitive actions based on the agent's actual position in the map are visible along with the corresponding graphs of maximum utility for the selected three actions.

11.5.3 Part 2 — Planning over the Hierarchy of Decision Spaces

At this stage of the experiment, the agent has a sufficient knowledge to fulfill the given task for the user. The user describes the goal state (lights on, doors opened),

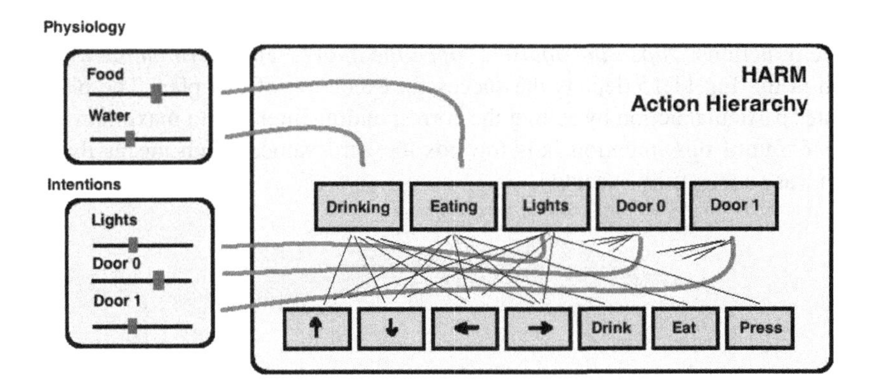

Fig. 11.13 Resulting hierarchy of actions. Drinking and eating is motivated by the agent's physiology, remaining actions are motivated by the agent's autonomously created intentions. The maximum level of action abstraction is one, therefore each abstract action is composed only of selected subset of primitive actions.

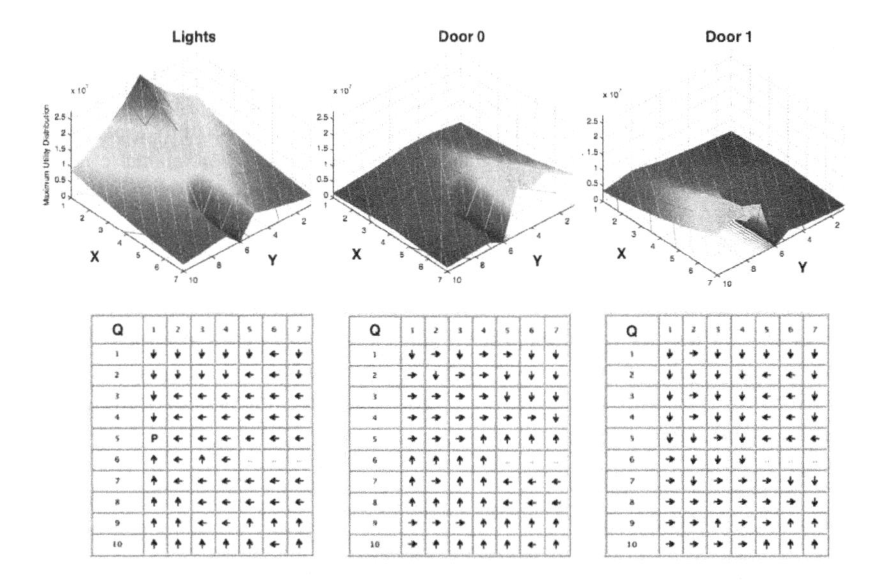

Fig. 11.14 The acquired primitive actions based on the agent's actual position in the map and the corresponding graphs of maximum utility for the selected three actions. Selected three behaviors, from the left are: *light control*, *door0 control* and *door1 control*. Note that lights are controlled by pressing the switch, but the doors are controlled just by approaching towards the door-switch position.

the agent creates the plan and executes it. The resulting plan is composed of sequence of actions: *"open the door0"*, *"open the door1"* and *"turn on the lights"*. Graph in the Fig.11.15 depicts the successful execution of this plan. The planner executes particular action by setting the corresponding intention to maximum value and waits until this intention falls towards the zero value, which means that the action was successfully executed.

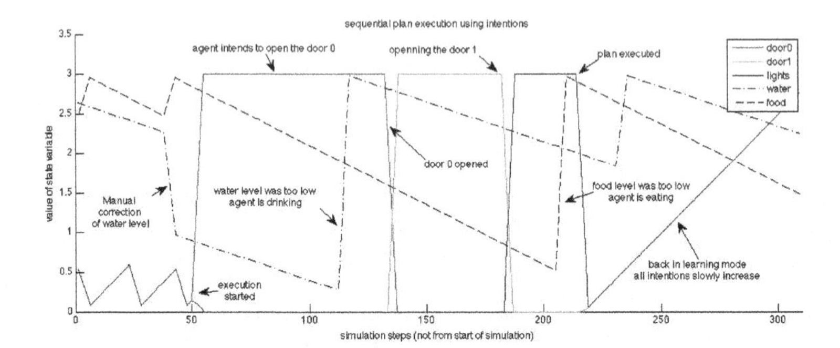

Fig. 11.15 Sequential execution of plan composed of sequence of actions: *"open door0"*, *"open the door1"*, *"turn on the lights"*. All intentions are set at the beginning to the zero. Afterwards, planner sets intentions to the maximum value in the same order as they are in the plan. The decrease of intention means that the action has been successfully executed. After this event the planner sets the new sub-gual. On the *Y* axis there is an amount of intention to execute particular action (dashed lines represent amounts of food and water in the agent's body), the *X* axis represents time.

From the graph we can observe another benefit of this approach: during the execution of plan the agent became thirsty and later hungry (dashed lines). At some point the agent's physiological need (agent's survival) became more urgent than the plan execution, therefore the agent temporarily interrupts the plan execution in order to drink and later to eat.

11.6 Conclusion

The goal of this project has not been simply to build machines that are similar to humans but to alter our perception of the potential capabilities of robots. Our current attitude toward intelligent robots, we assert, is simply a reflection of our own view of ourselves.

None of the mentioned approaches could handle the all of the comparable problems alone, therefore the main contribution of this architecture is in the combination of the advantages of particular subsystems and in their interconnection in such manner, that more complex behavior can emerge from their mutual interaction.

We believe that the main advantage of our architecture is in combination of hierarchical reinforcement learning subsystem and the planning engine in a way that their collaboration enables hierarchical planning while maintaining the domain independence. This is because the fact that architecture configures itself based only on the knowledge autonomously gained from the particular domain. This means that compared to widely used planners like in *Belief-Desire-Intention* architecture (BDI) [18], our agent does not need the predefined plan library - that is a priory knowledge. Recently, similar architectures, which try to combine planning and reinforcement learning (that is subset of our work) were found [19], [20]. The first architecture however has the similar disadvantage to the BDI mentioned above, and the second one uses a slightly approach for problem solving.

Based on our completed experiments in comparison with other approaches we believe that our architecture is superior to several widely-used principles in to days field of AI. Where the classical planners fail and the hierarchical ones need help of designer, our agent is still able to operate without bigger problems.

Acknowledgements. This research has been funded by the Dept. of Cybernetics, Faculty of Electrical Engineering, Czech Technical University in Prague and Centre for Applied Cybernetics under Project 1M0567.

References

1. Horáková, J., Kelemen, J.: From Rossums Universal Robots Toward Post-Human. In: Cybernetics and Systems, vol. 2, pp. 774–779. Austrian Society for Cybernetic Studies, Austria (2004)
2. Kelemen, J.: Myslenie a stroj, nakladatelstvo Kalligram, Bratislava, p. 384 (2010)
3. Husbands, P., Holland, O., Wheeler, M. (eds.): The Mechanical Mind in History. MIT Press (2008)
4. Steels, L., Brooks, R.: The artificial life route to artificial intelligence: Building Situated Embodied Agents. Lawrence Erlbaum Ass., New Haven (1994)
5. Wooldridge, M.R.: An Introduction to Multi-Agent Systems. John Wiley & Sons, New York (2002)
6. Albus, J.S.: A New Approach to Manipulator Control: the Cerebellar Model Articulation Controller (CMAC). Trans. ASME, Series G. Journal of Dynamic Systems, Measurement and Control 97, 220–233 (1975)
7. Handelman, D.A., Lane, S.H., Gelfand, J.J.: Integrating neural networks and knowledge-based systems for intelligent robotic control. IEEE Control Systems Magazine 10, 77–87 (1990)
8. Ryan, M., Pendrith, M.: RL-TOPs: An Architecture for Modularity and Re-Use in Reinforcement Learning. In: Proceedings of the Fifteenth International Conference on Machine Learning, pp. 481–487 (1998)
9. Nilsson, N.J.: Teleo-reactive programs for agent control. Journal of Artificial Intelligence Research 1, 139–158 (1994)
10. Vítků, J.: An Artificial Creature Capable of Learning from Experience in Order to Fulfill More Complex Tasks, CTU in Prague, FEE, Diploma thesis supervised by Nahodil, P., p. 142 (2011)
11. Kadleček, D.: Motivation Driven Reinforcement Learning and Automatic Creation of Behavior Hierarchies, CTU in Prague, FEE, PhD thesis supervised by Nahodil, P., p.143 (2008)

12. Kadleček, D., Nahodil, P.: New Hybrid Architecture in Artificial Life Simulation. In: Kelemen, J., Sosík, P. (eds.) ECAL 2001. LNCS (LNAI), vol. 2159, pp. 143–146. Springer, Heidelberg (2001)
13. Nahodil, P., Kadleček, D.: Adopting Animal Concepts in Hierarchical Reinforcement Learning and Control of Intelligent Agents. In: Proc. of 2nd IEEE/RAS-EMBS Intern. Conf. on Biomedical Robotics and Biomechatronics, BioRob, Scottsdale, pp. 122–131 (2008)
14. Grand, S.: Creation: Life and How to Make It, p. 230. Harvard University Press (2003)
15. Fikes, R., Nilsson, N.: STRIPS: A New Approach to the Application of Theorem Proving to Problem Solving. Artificial Intelligence 2, 189–208 (1971)
16. Sacerdoti, E.D.: Planning in a Hierarchy of Abstraction Spaces. Artificial Intelligence 5(2), 115–135 (1974)
17. Erol, K., Nau, D., Hendler, J.: HTN Planning: Complexity and Expressivity. In: Proc. National Conference on Artificial Intelligence (AAAI 1994), pp. 1123–1128. MIT Press, Seattle (1994)
18. Sardina, S., de Silva, L., Padgham, L.: Hierarchical Planning in BDI Agent Programming Languages: a Formal Approach. In: AAMAS 2006 Proceedings of the Fifth International Joint Conference on Autonomous Agents and Multiagent Systems, pp. 1001–1008. ACM Press, NY (2006)
19. Ryan, M., Pendrith, M.: RL-TOPs: An Architecture for Modularity and Re-Use in Reinforcement Learning. In: Proc. of the 15th International Conference on Machine Learning, pp. 481–487 (1998)
20. Ross, M.: Hierarchical Reinforcement Learning: A Hybrid Approach, PhD Thesis, The UNI of New South Wales, School of Computer Science and Engineering (2002)

Part III
When Artificial Becomes Natural

The way from artificial to natural is extremely complex, yet often surprisingly fast. Let us listen to our intuition and pretend that we know what it really is to be "artificial", "natural", and what it is to be "intelligent". Even from this intuitive stance we can see that systems of artificial intelligence have already infiltrated our natural environment and that we can link them with our minds, bodies and behaviour without losing anything of our naturalness as humans. We can use AI to enhance our body—either when the body is failing, or even to extend its limits. We can use AI to enhance our mind—to form extended mind. And we can use AI to enhance our environment and create its ambient intelligence. Utilising new technologies, we can externalise our cognitive functions and even our embodiment.

Chapter 12
A View on Co-existence of Various Entities in Intelligent Environments

Peter Mikulecky and Petr Tucnik

Abstract. There is a number of traditional points of view from which intelligent environments are usually investigated. The most frequent among them are technological, social, economical, ethical, or political point of view. However, up to now there were just a few papers devoted to a research focused on co-existence of various entities that share such an intelligent environment and have to interact there. If we are viewing an intelligent environment as a collection (if not a community) of intelligent entities, capable of communication and performing activities based on a kind of mutual co-operation, then a kind of a co-existential point of view could be useful for further contemplations about various issues arising from this co-existence. The purpose of this paper is to present and discuss important aspects of co-existence of intelligent entities of various types (including humans) in intelligent environments and to formulate some interesting and important problems related to such a co-existence.

12.1 Introduction

Intelligent environments, based on broad exploitation of nowadays technologies such as smart sensor networks, ambient intelligence, or other similar approaches, are rapidly developing with a plenty of interesting applications. Such intelligent environments are usually investigated from traditional points of view: *technological, social, economical, ethical*, and even *political.*

Within the *technological point of view* it is natural to study and develop technical devices, information, knowledge and communication technologies which will make the implementation of the vision of ambient intelligence possible. Key technologies might, among others, include also knowledge management, artificial

Peter Mikulecky · Petr Tucnik
Department of Information Technologies, University of Hradec Králové,
Rokitanského 62, 500 03 Hradec Králové, Czech Republic
e-mail: {peter.mikulecky,petr.tucnik}@uhk.cz

J. Kelemen et al. (Eds.): Beyond Artificial Intelligence, TIEI 4, pp. 183–195.
DOI: 10.1007/978-3-642-34422-0_12 © Springer-Verlag Berlin Heidelberg 2013

intelligence, user interfaces, communication and network services, as well as solution to the problems of security and protection of data and information.

The *social point of view* focuses on studying the influences of social, economic and geopolitical trends on the quality of everyday life and the acceptance of using solutions employing information and communication technology (ICT) for solving problems accelerated by the above mentioned trends. Areas of problems of more global nature include, for example, ageing of society, multicultural society, lifelong education, the problem of consumer society, globalization, etc.

The *economical point of view* follows the fact, that the concept of AmI is strongly motivated also by economic aspects—probably economic motivation is the most significant incentive in this area and a standpoint that has not been deeper studied up to now. A discussion about real time or now-economy has been presented by Bohn and others [3], where more and more entities in the economic process, such as goods, factories, and vehicles, are being enhanced with comprehensive methods of monitoring and information extraction.

The *ethical point of view* was mentioned already in [13] or again in [3]. However, this important area deserves much more attention as it attracted up to now. There are important issues to be addressed regarding privacy, handling personal information, monitoring living space of persons, their conversations, etc. These are issues representing most problematic concerns for potential customers and users of intelligent environments.

The *political point of view* has its starting point in the resolution adopted at the Lisbon congress of the EU in 2000, on the basis of which the European Commission resolved to secure Europe's leading role in the field of generic and applied technologies for creation of knowledge society, and thus increase Europe's ability to compete successfully, and enable all European citizens to take advantage of the merits of knowledge society. To this effect, the new technologies must not be the cause for excluding some groups of citizens from society, but must ensure universal and equal approach to its—both digital and therefore also knowledge—sources.

However, up to now there were just a few papers devoted to research focused on co-existence of various entities that share such an intelligent environment and that have to interact there. Some contemplations in this direction can be found e.g. in [3] or in [11].

If we are viewing an intelligent environment as a collection (if not a community) of intelligent entities, capable of communication and performing activities based on a kind of mutual coordination, then a kind of *co-existential point of view* could be useful for further contemplations about various issues and problems arising from this co-existence. If we consider the effect of introducing Ambient Intelligence technologies and approaches in an environment, we have to take into consideration that what we are creating in such a case is an environment based on communicating intelligent entities.

In the case of interacting living organisms sharing a populated environment, we certainly can use the term community as it is usually used in biology or other areas. Analogically, taking into account the nature of intelligent environments where both living and artificial entities are sharing the populated environment and interacting mutually, such intelligent entities create in a sense a community as well. Such a community is supposed to communicate throughout the environment and accomplish a number of activities aiming to support respective human activities. In other words, in such an intelligent environment we have to take into account a number of aspects of mutual co-existence of human beings with these intelligent entities, or more precisely, with a kind of a community of these entities.

The purpose of our paper is to present some important aspects of co-existence of intelligent entities of various types (including human beings) in intelligent environments and to formulate some important problems that arise from such a co-existence.

12.2 Related Works

Indeed, the idea of ambient intelligence grew on achievements in various related areas, ubiquitous computing, pervasive computing and artificial intelligence being most important among them. As Cook, Augusto and Jakkula [8] pointed out, the basic idea behind AmI is that by enriching an environment with technology (e.g., sensors and devices interconnected through a network), a system can be built such that it acts as an "electronic butler", which senses features of the users and their environment, then reasons about the accumulated data, and finally selects actions to take that will benefit the users in the environment.

There is a number of interesting papers dealing with certain, but prevailingly particular aspects of co-existence of various entities in environments with ambient intelligence (or shortly, intelligent environments). The majority of them are focused on smart home environments, however, for the sake of our research the most interesting are those modelling the environment via multi-agent systems. Let us mention here as examples the papers devoted to the MavHome smart home project [6], location-awareness in smart homes [15], or earlier, but still actual paper [9] formulating seven challenges to be taken into account in the process of smart homes vision realization. An interesting and rather exhaustive survey of the smart homes area can be found in [5], while recent survey [4] focuses on a higher degree of context-awareness, the situation-awareness.

In what follows, we shall mention a couple of important papers closely related to our theme of co-existence of various intelligent entities in an intelligent environment. An alternative approach to designing ambient intelligent environments is introduced in [1], using a multi-agent system consisting of agents that represent inhabitants (humans, animals, plants, and objects) of the environment and physical devices (sensors and actuators) that control and monitor the environment. In this approach, the inhabitants are able to compromise their own needs for the betterment of the environment as a whole. This synergy creates a balance where each inhabitant potentially receives sub-optimal environmental conditions but the environment as a whole achieves an optimal level.

The ambient intelligent environment is represented here by a multi-agent system consisting of agents (problem-solvers capable of functioning effectively and efficiently in complex and dynamic environments) that represent inhabitants of the environment (humans, animals, plants, and inanimate objects) as well as physical devices (sensors and actuators) that control and monitor the environment. Using profile modelling, each agent is designed to model the biological or basic needs of an inhabitant of the environment. Each inhabitant's environmental needs motivate it to plan and schedule events in the environment by searching for a set of environmental settings that allow the inhabitant to preserve its well-being and physical structure. This search is done using a genetic algorithm with the agents' feedback.

The approach used in [1] differs from that traditionally used by researchers in the area of ambient intelligent environments where the user is always in charge of the environment. This is known as the *user is king* axiom and it currently dominates the ambient intelligent environment community. According to Becerra and Kremer [1], the *user is king* axiom is not always true, for a person would not invest money in purchasing a bonsai, an orchid or a valuable painting if he/she is unwilling to take care of the item by properly maintaining it. An ambient intelligent environment that follows this axiom presents a simplistic behaviour that, for the most part, ignores other entities and only reacts to its human user.

Instead, the user should be expected to take into account also the needs of other entities co-existing with him/her in the environment and to balance his/her needs with the other entities' needs. In other words, the work [1] advances the current research by allowing ambient intelligent environments to take into account the environmental needs of those other inhabitants that are important but can't interact with the environment. Another important contribution is the inclusion of more than one human in the decision making process.

If we wish to perform research focused on certain aspects of the co-existence of various intelligent entities (human and non-human, natural or artificial) in an intelligent environment, we cannot neglect problems related to introducing more social intelligence into such intelligent environments. The first papers focused on this were [10] or [3]. According to Markopoulos and his colleagues [10], we are increasingly led to anticipate that ambient intelligent technology will mediate, permeate, and become an inseparable component of our everyday social interactions at work or at leisure. Therefore, the social dimension of ambient intelligence is clearly an important topic for human-computer interaction research.

In relation to the social component of ambient intelligence Markopoulos and his colleagues formulated four critical challenges to human computer interaction research [10]:

- Designing ambient intelligence systems and environments so that they can be perceived as socially intelligent.
- Designing intelligence that will support human-to-human cooperation and social interactions.
- Finding the ways how to evaluate social intelligence, that is, to be able to verify that one design of an intelligent environment is superior to another with regards to how they are perceived as socially intelligent.

- Finding the benefits of social intelligence, or, in other words, to verify the relevance of a research program on social intelligence we first need to establish that it does provide some added value for users of ambient intelligent systems.

Even today we can agree with one of the Markopoulos et al. [10] conclusions that a theoretical framework for social intelligence suitable for the study of ambient intelligence environments still needs to be developed. We still do not know any fundamental progress in this direction, although some partially interesting results or at least contemplations can be found e.g. in the paper by Nijholt [14] or earlier [13].

Another important group of problems related to our co-existential point of view seems to be based on the fact that multiple users sharing the same intelligent environment have to be taken into account. If we consider the situation of multiple users sharing the same intelligent environment, two essential cases are possible:

- All the users are human and we cover in our model just these as those who actively influence the environment; or
- Not all the "users" are human (some of them could be animals, plants, artificial entities, etc.) and we should consider both actively and passively influencing the environment.

The most interesting related papers here are those focused on solutions of balancing preferences of the users to be a compromise beneficial for all in the environment, see again [1] as a good example of this line of research. The further important papers we should mention in this relation are papers merely oriented on finding a solution to the problem of multiple users in an intelligent environment, e.g., [7], [12], or [16]. In the last paper by Roy et al. [16], an important result has been proven, that the problem of maximizing the number of successful predictions of multiple inhabitants' locations in a smart home is NP-hard. In other words, it is computationally infeasible to find an optimal strategy for maximizing the number of successful location predictions across multiple inhabitants of a smart environment. Roy and his colleagues [16] then devised a suboptimal solution based on game theory that attempted to reach equilibrium and maximized the number of successful predictions across all inhabitants.

According to [7], under the hypothesis that each inhabitant in a smart environment behaves selfishly to fulfil his/her own preferences or objectives and to maximize his/her utility (see also Chap. 7 of this book), the residence of multiple inhabitants with varying preferences might lead to conflicting goals. Once again, this is a result of the *user is king* principle. Under this circumstance, a smart environment must be intelligent enough to strike a balance between multiple preferences, eventually attaining an equilibrium state. Or, we have to admit that users' approach must be changed in the sense of [1], that is, each user (and the humans are here in first line) have to take into account also the other users' needs and requirements leading to a compromise solution of the conflicting goals problem.

12.3 Co-existential View

The co-existential view should be focused on the problem arising from the relatively simple fact that various information devices integrated into people's everyday life represented with their intelligent interfaces capable to communicate with people, can be understood as relatively independent entities with certain degree of intelligence. Their intelligence varies, of course, from rather simple level of one-purpose machines to relatively intelligent and complex systems (e.g., an intelligent building, or an intelligent vehicle). These intelligent entities co-operate one with another, and all of them from time to time have to co-operate with humans.

Considering humans as another intelligent entity, we are able to study the co-existence of various intelligent entities in real world. This will lead to an investigation of a number of different interesting aspects of such a co-existence, and also to a number of potentially important consequences for human lives (cf. Mateusz Woźniak's contribution in Chap. 16 of the present volume).

When taking into account such artificial entities with a certain degree of intelligence and with a mechanism for the initiation of its activity, where the activity is oriented on certain benefit (or service) to human beings, we are able to investigate the following basic problems related to them:

- Various types or levels of such artificial entities;
- Their mutual relationships as well as their relationships with humans;
- Their communities (virtual as well as non-virtual);
- Their co-existence, collaboration and possible common interests;
- Their co-existence and collaboration with humans;
- Antagonism of their and human interests;
- Ethical aspects of the previous problems, etc.

The first impression from the AmI idea is that humans are surrounded by an environment in which there are microprocessors embedded in any type of objects—in furniture, kitchen machines (refrigerator, coffee maker, etc.), other machines (e.g., washing machine, etc.), clothing, toys, and so on. Of course it is depending on the type of the particular environment, there are clear differences between an environment in a hospital when compared with a luxurious private house, or in comparison with a university environment.

It is straightforward that when speaking about intelligent artificial entities capable of mutual communication, we could certainly expect some relatively intelligent behaviour of such a community. We can speak about the emergent behaviour of such a community that can be modelled by a multi-agent system, serving to some purpose which is considered to be beneficial for humans. However, the emergent behaviour of such an artificial community can be potentially dangerous—if the possible goal of the community differs from the human interests, or if the community is simply unable to serve the human being goals for various (maybe also technical) reasons. We certainly have to take into account such questions, like:

- How to tune all the emergent behaviour of the particular environment to be able to serve the particular human being goals?

- What to do if the emergent behaviour of the environment is not in accord with the human aims, or even if it is contradictory to the intentions of the particular human?
- How the privacy of a particular human will be respected in an intelligent environment?
- Is the particular information about the concerned human safe from being exploited by another person?

Of course, these are just a few of possible questions which could arise in relation to the first attempts to introduce the AmI idea into the life. Some of other issues certainly will be mentioned in the future.

12.4 Perceiving Users and Environment

When considering implementation of ambient environment, one of most important factors to be considered is how will be such system perceived by the user. There are two points of view, shown in the Fig. 12.1 and Fig. 12.2.

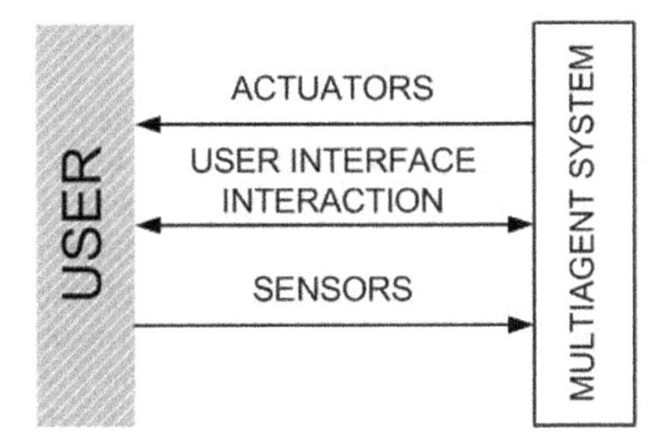

Fig. 12.1 How the user is perceived by the system

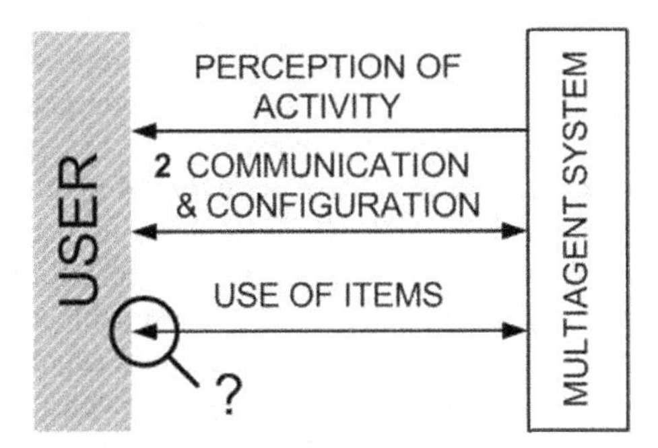

Fig. 12.2 How the system is perceived by the user

In this part of the text, we will focus first on how the system perceives a user (Fig. 12.1). At this point, it will be mostly home environment we will take into consideration here.

There are many different potential users in the environment. Their difference lies in their ability to utilize environment features (e.g. appliances, terminals, controls), objective risks involved in it (e.g. children are not allowed to use everything in the kitchen), health condition (e.g. disabled person) or simply their relation to given area (e.g. inhabitant, visitor, intruder). Since there are many problems arising from contradictory goals and needs of different users, it is useful to use concept of roles in the system.

Roles can be useful in many ways. It is approach that is easy to implement. Each person has assigned role corresponding with their needs and reflecting their status in the given environment. However, there are many problems to be solved in real-world applications. Role of an individual in a social group continuously changes over time—people are getting older, sick, married, etc. and assigned role in environment should reflect that.

The role is defining formal frame of bi-directional relationship between intelligent environment and its inhabitant. User has certain privileges, but also limits in how the system may be used. The system, on the other hand, has an obligation to be helpful in fulfilling user's needs but cannot act too much proactively in order to remain discrete and non-intrusive.

Problem with perceiving users in environment is in their variability and differences. One way to approach this problem is following. When we use multi-agent system, where every agent represents individual appliance or object in environment capable of some type of activity (and using such multi-agent idea for representing ambient environment is a very frequent approach), it is possible to use similar approach to include different types of users into system. A notion of *"user"* can be extended to include humans as well as animals, plants and even inanimate objects such as valuable paintings or decorations. When it is considered thoroughly, all these groups of users are similar to each other in at least one key aspect—there exist their specific needs which the system is supposed to take care of. Obviously, each above mentioned category of users is manifesting different form of behaviour and needs, but in principle their relationship with the system is similar. Using such approach will ensure that every requirement towards the system is represented somehow in a form of more or less important *"user"* and nothing is omitted. The roles are also useful here to clarify the priority of such request.

Other problem that could arise is hidden in overlap of both perspectives. While system perceives humans, animals, plants and objects as users (with individual needs and goals), environment itself is perceived as a collection of agents, working together and forming multi-agent system. It is reasonable to assume that many objects represented as agents (with a capability to perform actions according to plan created by decision-making component of multi-agent system) can be perceived as *"users"* as well, since they have their needs that must be fulfilled. This consequently results in double-sided representation of some objects in environment: on the one hand they are accepted as agents, fulfilling their role within the system; on the other hand they are recognized as users at the same time.

This opens many possible questions and problems for future work. Up to now, domain of intelligent environments and ambient intelligence lacks standardized formal structures. Although axiom *user is king* is slowly on withdraw, the need for incorporation of representational structures of larger scale rises.

The challenges in domain of ambient intelligence are nowadays much more demanding and complex than those of traditional artificial intelligence. Not only that it is expected that intelligent environment will be working efficiently and be able to plan and coordinate actions of many heterogeneous entities in one system (which is a "traditional" task). The system is also required to pursue goals that are continuously changing, reflecting more or less unpredictable user's needs. Moreover, there are many mutually conflicting interests at the same time to be taken into consideration and these conflicts must be resolved.

Given the continuously changing, unpredictable nature of real-world environments, intelligent environment should be able to manifest a great deal of adaptability. Future research in area of intelligent environments should be focused on multiple users: how to help them pursue their goals and fulfil their needs. As it was described above, multiple-user approach may be extended to cover even other entities, different from humans, with their specific needs.

However, the user perspective (Fig. 12.2) is more important here. User is capable to perceive the activity in his/her surroundings, but there is a strong tendency to make evaluation of system's performance based exclusively on such observation. The communication and configuration level represents requested and expected exchange of information here. Last layer of Fig. 12.2 is comprised of using appliances and items. What is to be considered and decided is appropriate level of interference with user's actions. While user is using items or appliances, system may or may not try to be helpful, and this would result in proactive behaviour. Depending on user's personal preferences or needs, this is interpreted as an aid or an interference.

In every ambient environment serving as a home or a workplace (like office, for example), it is user's satisfaction and comfort that matters the most. System's performance may be perceived in many ways, depending on individual preferences of the user. In general, the system's function is supposed to be discrete and interactions kept to minimum.

Let us focus briefly on how user perceives an environment with autonomous decision-making features. It is a marketing thing that ambient environment is sometimes presented as fully robotic and automated, depending heavily on artificial intelligence as its main component. To some limited extent this could be true, but, in reality, this may create false expectations on the part of the user. Similar situation already occurred in the past with misused term *"artificial intelligence"*. Many systems claimed to be *"intelligent"* while it was not entirely true and this is what in consequence led to devaluation of semantic value of this term. Nonetheless, when user is situated in the ambient environment, his way of perceiving is intensely influenced by his beforehand expectations.

In the well-designed system, many of the systems functions will remain hidden and user will be interacting with only a limited part of the system. In fact, environment takes over some decisions instead of the user, while still allowing him to interfere or adjust it if setting is perceived as unsatisfactory. This is usually related to energy consumption optimization, temperature, light intensity, humidity, pressure etc.—this set of features could be called environmental conditions. Keeping optimal environmental conditions consciously—manually may be better expression here—may be tedious and uncomfortable for the user. His interference is always possible, but not required. System is able to function independently, changing environmental conditions continuously to the optimal level. For example, if the user is not in the room for a long time, system may decrease heating, turn off lights, etc. This is easing user's workload, increasing his comfort while remaining in such environment.

Many authors, e.g. [4], [1] or [2] and others focus on functioning features of the ambient environment which are only loosely related to the user, there is no direct connection in this aspect, but they are rather focused on satisfying needs of other entities in the environment. Becerra [2] specifically mentions search for ideal conditions for animals, valuable decorative items, plants, humans—all at the same time. He proposes genetic algorithms to be appropriate tools for finding optimal configuration.

Environmental conditions are only one of many possible examples of user-system interaction. But here is easy to see important trend in AmI—ambient systems are increasingly focused on optimization of its behaviour on general level, satisfying needs of everyone and everything in them at the same time. This creates conflicts that have to be automatically resolved.

12.5 Significance of Roles

Among important further aspects of the co-existence of humans with various intelligent entities certainly counts the problem of role assignment. Roles, already mentioned in the previous part, could possibly serve as a good concept for users' classification with respect to the environment in which they have to perform their activities. However, a number of possible issues with roles can certainly be identified:

- Wrong role assignment to a particular person;
- Shifts in the case of long-time roles (a child is growing and becoming an adult, etc.);
- Originally strange person becomes a relative and therefore starts fulfilling of a completely different role with new permissions;
- Changes in permissions related to various functionalities of the system;
- Permissions assigned to various roles can be in a conflict in certain situations (e.g., a husband and his wife);
- Etc.

When living in an environment that claims to be *"intelligent"* in some way, one rightfully expects it to be helpful. Problem is that priorities and desired behaviour of any person change over time. In this part of the text, term *"user"* will be used again in its traditional meaning, referring to a human.

For healthy, adult person, effectiveness of the system is measured mainly by level of comfort it can provide. In a simplified perspective, the system should provide optimal living and/or working conditions, discrete help with minor tasks and take care of some responsibilities automatically. Interaction with system should be kept at minimum level, as well as proactive behaviour of the system towards the user (Fig. 12.3).

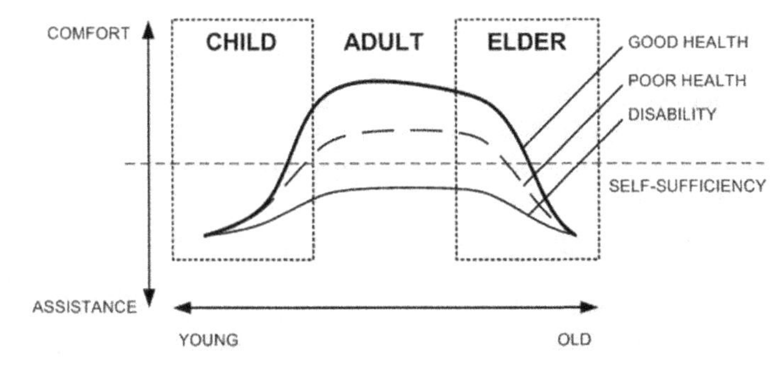

Fig. 12.3 Example of a stimuli changing user's priorities. Expected role of system is depending on health condition and age. Self-sufficiency level is indicating ability of person to take care of his/her individual needs without help of others.

Situation is different for disabled person. Here, the effectiveness of a system is measured by variety and quality of an assistance it can provide. Proactive behaviour or more frequent interactions are not a problem here, since it helps the system to be more useful. It is also reasonable to anticipate that roles with higher priority will be assigned to users which are more dependent on the system, i.e. are situated lower on the Fig. 12.3.

There are also many interesting problems that seem to be unsolved up to now that are strongly related to the concept of roles and permissions assigned to these roles. Some of them are listed here:

- Various humans have various preferences in their environment functionalities exploitation, how to cope with that?
- Questions of ergonomics, but people like to have their thinks ordered in accord with their taste.
- Animals in the environment, domestic animals, but also other, coming from outside (birds or even insects).
- Non-specified persons (without a role) coming into the environment—are they always a source of potential threat?

In any case, role is a useful concept for rationalization of behaviour of AI system towards the user, and it deserves to be evolved further on.

12.6 Conclusions

As Bohn with colleagues [3] pointed out, the fundamental paradigm of ambient intelligence, namely the notion of disappearing computer (computers disappear from the user's consciousness and recede into the background), is sometimes seen as an attempt to have technology infiltrate everyday life unnoticed by the general public in order to circumvent any possible social resistance. However, the social acceptance of ambient intelligence will depend on various issues, sometimes almost philosophical ones. The most important issue seems to be our changing relationship with our environment. Probably we have to change our old living habits in favour of new, more appropriate ones.

Main idea of this text was to stress some interesting open problems which emerge as the research of intelligent environments evolves. The approach focused on single-user case can be considered obsolete as this is quite rare scenario in real world applications. In order to design ambient systems fully utilizing their potential (with respect to users' privacy and preferences), more attention in our future work will be focused on multiple-user environments, resolving conflicts, concept of roles and human-computer interactions. We should anticipate sensibly that our living environments would play much more important roles in the future when compared to present days.

Acknowledgements. This work and contribution were supported by the project of the Czech Science Foundation No. P403/10/1310 "SMEW - Smart Environment at Workplaces" and the Specific Research Project IPV 09/2011 "AmIRRA 2 - Ambient Intelligence Related Research Activities, Part 2".

References

1. Becerra, G., Kremer, R.: Ambient Intelligent Environments and Environmental Decisions via Agent-based Systems. Journal of Ambient Intelligence and Humanized Computing 2(3), 185–200 (2011)
2. Becerra, G.: Environmental Decisions in Ambient Intelligent Environments. MSc Thesis, Department of Computer Science, The University of Calgary (2011)
3. Bohn, J., Coroamă, V., Langheinrich, M., Mattern, F., Rohs, M.: Social, Economic, and Ethical Implications of Ambient Intelligence and Ubiquitous Computing. In: Weber, W., Rabaey, J., Aarts, E.H.L. (eds.) Ambient Intelligence, pp. 5–29. Springer, Heidelberg (2005)
4. Buchmayr, M., Kurschl, W.: A Survey on Situation-Aware Ambient Intelligence Systems. Journal of Ambient Intelligence and Humanized Computing 2(3), 175–183 (2011)
5. Chan, M., Estève, D., Escriba, F., Campo, E.: A Review of Smart Homes-Present State and Future Challenges. Computer Methods and Programs in Biomedicine 91(1), 55–81 (2008)

6. Cook, D.J., Youngblood, G.M., Das, S.K.: A Multi-agent Approach to Controlling a Smart Environment. In: Augusto, J.C., Nugent, C.D. (eds.) Designing Smart Homes. LNCS (LNAI), vol. 4008, pp. 165–182. Springer, Heidelberg (2006)
7. Cook, D.J., Das, S.K.: How Smart Are Our Environments? An Updated Look at the State of the Art. Pervasive and Mobile Computing 3(2), 53–73 (2007)
8. Cook, D.J., Augusto, J.C., Jakkula, V.R.: Ambient Intelligence: Technologies, Applications, and Opportunities. Pervasive and Mobile Computing 5(4), 277–298 (2009)
9. Edwards, W.K., Grinter, R.E.: At Home with Ubiquitous Computing: Seven Challenges. In: UbiComp 2001:Proceedings of the 3rd ACM International Conference on Ubiquitous Computing, pp. 256–272 (2001)
10. Markopoulos, P., de Ruyter, B., Privender, S., van Breemen, A.: Case Study: Bringing Social Intelligence into Home Dialogue Systems. ACM Interactions 12(4), 37–44 (2005)
11. Mikulecký, P.: Remarks on Ubiquitous Intelligent Supportive Spaces. In: Jegdic, K., Simeonov, P., Zafiris, V. (eds.) Proc. of the 15th American Conference on Applied Mathematics and Proc. of the International Conference on Comp. and Information Sciences, vol. I, II, pp. 523–528. WSEAS Press, Athens (2009)
12. Muñoz, A., Botía, J., Augusto, J.C.: Intelligent Decision-Making for a Smart Home Environment with Multiple Occupants. In: Ruan, D. (ed.) Computational Intelligence in Complex Decision Systems, pp. 329–375. Atlantic Press (2010)
13. Nijholt, A., Rist, T., Tuijnenbreijer, K.: Lost in Ambient Intelligence? In: CHI 2004: Proceedings of the ACM Conference on Computer Human Interaction, pp. 1725–1726 (2004)
14. Nijholt, A.: Google Home: Experience, Support and Re-experience of Social Home Activities. Information Sciences 178(3), 612–630 (2008)
15. Roy, A., Das, S.K., Basu, K.: A Predictive Framework for Location-Aware Resource Management in Smart Homes. IEEE Transactions on Mobile Computing 6(11), 1270–1283 (2007)
16. Roy, N., Roy, A., Das, S.K.: Context-Aware Resource Management in Multi-Inhabitant Smart Homes: A Nash H-Learning based Approach. In: PERCOM 2006: Proceedings of the 4th Annual IEEE International Conference on Pervasive Computing and Communications, pp. 158–169 (2006)

Chapter 13
Multi-agent Systems in Industry: Current Trends & Future Challenges

Paulo Leitao

Abstract. This paper introduces the multi-agent systems paradigm and presents some industrial applications of this AI approach, namely in manufacturing, handling and logistics domains. The road-blockers for the current weak adoption of this technology in industry are also discussed, and finally the current trends and several future challenges are pointed out to increase the wider dissemination and acceptance of the multi-agent technology in industry.

13.1 Introduction

The application of artificial intelligence (AI) mechanisms allows the development of intelligent machines/systems capable to solve very complex engineering problems. Multi-agent systems is one paradigm, derived from the distributed artificial intelligence and artificial life fields, that allows an alternative way to design distributed control systems based on autonomous and cooperative agents, exhibiting modularity, robustness, flexibility, adaptability and re-configurability. This paper introduces the multi-agent systems paradigm and presents some industrial applications of this AI approach, namely in manufacturing, handling and logistics domains. The road-blockers for the current weak adoption of this technology in industry are also discussed, and finally the current trends and several future challenges are pointed out to increase the wider dissemination and acceptance of the multi-agent technology in industry.

Paulo Leitao
Polytechnic Institute of Bragança, Campus Sta Apolonia, Apartado 1134,
5301-857 Bragança, Portugal
Artificial Intelligence and Computer Science Laboratory, R. Campo Alegre 102,
4169-007 Porto, Portugal
e-mail: pleitao@ipb.pt

J. Kelemen et al. (Eds.): Beyond Artificial Intelligence, TIEI 4, pp. 197–201.
DOI: 10.1007/978-3-642-34422-0_13 © Springer-Verlag Berlin Heidelberg 2013

13.2 Artificial Intelligence and Multi-agent Systems

The management of complexity, currently found in systems ranging from washing machines to Airbus A380 aircrafts, requires the use of proper mechanisms and techniques. Artificial intelligence (AI), introduced by John McCarthy in 1956, is the science and engineering of making intelligent machines, especially intelligent computer programs mimicking the human though [7]. AI is becoming an essential part of the technology industry, providing solutions for several complex problems in engineering and computer science, namely:

- Game playing, e.g. machines beating human chess players.
- Optimization, e.g. optimizing logistics and production processes.
- Pattern recognition, e.g. detection of trends and patterns in medical or production diagnosis.
- Computer vision, e.g. the navigation of autonomous mobile robots and analysis of medical images.
- Speech recognition, e.g. supporting human-machine interfaces.
- Intelligent control, e.g. providing adaptive and intelligent behaviour to control processes.

When applying AI techniques, several topics should be considered, namely the perception, reasoning, knowledge, planning and learning, as well some philosophical issues about the ethics of creating artificial intelligent beings.

The multi-agent systems (MAS) [10][1] is a paradigm that takes inspiration from several disciplines, mainly from distributed artificial intelligence (DAI) and artificial life (that is related to study and model systems possessing life, i.e. capable of reproducing, surviving and adapting in hostile environments). Multi-agent systems are based on a society of distributed autonomous, cooperative entities, each one having a proper role, knowledge and skills, and a local view of the world, being its behaviour regulated by simple rules. Agent-based solutions replace the centralized, rigid and monolithic control by a distributed functioning where the interactions among individuals lead to the emergence of "intelligent" global behaviour (see Fig. 1). Note that such systems exhibit high degree of autonomy and re-configurability, without a fixed client-server structure.

MAS is aligned with the current trend to build modular, intelligent and distributed control systems, which exhibit innovative features, like the agile response to the occurrence of disturbances and the dynamic re-configuration on the fly, i.e. without the need to stop, re-program and restart the process.

13.3 Applications of MAS in Industry

The MAS approach is suitable to support the current requirements for modern control systems in industrial domains, providing flexibility, robustness, scalability, adaptability, re-configurability and productivity. MAS is being applied with success to a wide range of domains, namely electronic commerce, graphics (e.g., computer

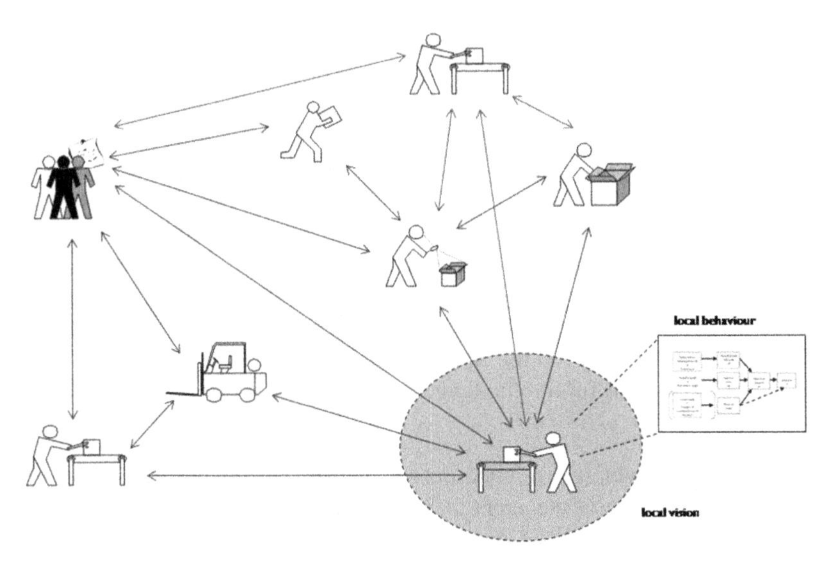

Fig. 13.1 MAS working in practice

games and movies), transportation, logistics, robotics, manufacturing, telecommunications and energy.

As examples, it is possible to refer the application of multi-agent systems solutions in the Daimler Chrysler factory of engines in Stuttgart [9], Tankers International that operates one of the largest oil tanker pools in the world [3], Air Liquide America to optimize the distribution of medical and industrial gases [2] and US Navy ships to control the heating, ventilation and air conditioned (HVAC) systems [5]. A deep analysis of industrial applications of MAS can be found in [4] and [8].

The analysis of the surveyed industrial applications of agent-based solutions allows extracting the following conclusions:

- Relatively small adoption of agents in industry, being the implemented applications limited in terms of functionality.
- The solutions address mainly the high-level control or the pure software systems (e.g. the electronic commerce).
- Little enthusiasm from both the technology providers and the industry companies.

The reasons for this weak adoption in industry were already widely discussed in the literature by several authors, namely [4] and [6]. Briefly, the main road-blockers are the required initial investment, the need to adopt the distributed thinking, the interoperability in distributed heterogeneous systems, the missing standardization, the real-time constraints and the missing technology maturity.

13.4 Current Trends and Future Challenges

Lately, some promising perspectives for the adoption of the agent technology were provided by the development of multi-agent based solutions by several software developers companies, e.g. NuTech Solutions, Magenta Technology, Smart Solutions and Whitestein Technologies, and by several automation technology providers, e.g. Rockwell Automation and Schneider Electric. However, the main trend in the industrial application of multi-agent systems is to convince industry people of the benefits of using agents, e.g. by providing demonstrators running in industry that shows the maturity, flexibility and robustness of agent-based solutions. This will allow industrial companies to "believe" in the agent technology and its principles.

Additionally, several future challenges can be pointed out in industrial agents, namely:

- *Standardization*, which is pointed out by industry as a major challenge for the industrial acceptance of the agent technology, since standards may affect the development of industrial MAS solutions, namely the IEEE FIPA (Foundation for Intelligent Physical Agents), IEC 61131-3, IEC 61499, ISA 95 and semantics and ontologies standards.
- *Integration of other complementary technologies*, e.g. IEC61131-3 and IEC 61499 approaches to implement the low-level control that is not addressed by the agents, and Service Oriented Architectures (SOA) / Web services to solve the interoperability problems allowing the vertical and horizontal integration.
- *Mature engineering development methodologies, deployment and tools*, that simplifies the engineering of agent-based systems. For this purpose, simulation is a need to test the emergent behaviour before the real deployment.
- *Bio-inspired techniques*, to enhance the engineering of more robust, adaptive, reconfigurable and responsive systems. In particular, self-organization is mandatory to support re-configuration and evolution, being also important to consider other self-* properties, such as self-learning, self-adaptation, self-optimization and self-healing.

The fulfilment of these challenges leads to the development of more powerful agent-based systems that may be better accepted by industry.

13.5 Conclusions

As conclusions, AI provides a set of advantages to improve the performance of automatic complex systems, and the multi-agent systems, as a paradigm derived from AI, is suitable to address the current requirements imposed to industrial companies. In spite of being already adopted in several industrial domains, the multi-agent technology still has a long and difficult path to be traversed for a wider acceptance of these AI concepts in industry.

References

1. Ferber, J.: Multi-Agent Systems: An Introduction to Distributed Artificial Intelligence. Addison-Wesley Professional (1999)
2. Harper, C.N., Davis, L.: Evolutionary computation at American Air Liquide. In: Evolutionary Computation in Practice, pp. 313–317 (2008)
3. Himoff, J., Skobelev, P., Wooldridge, M.: Magenta technology: multi-agent systems for industrial logistics. In: AAMAS 2005 Proceedings of the Fourth International Joint Conference on Autonomous Agents and Multiagent Systems, pp. 60–66 (2005)
4. Leitão, P.: Agent-based distributed manufacturing control: A state-of-the-art survey. Engineering Applications of Artificial Intelligence 22(7), 979–991 (2009)
5. Maturana, F.P., Staron, R.J., Hall, K.H., Tichý, P., Šlechta, P., Mařík, V.: An intelligent agent validation architecture for distributed manufacturing organizations. In: Camarinha-Matos, L. (ed.) Emerging Solutions for Future Manufacturing Systems, pp. 81–90 (2004)
6. Mařík, V., McFarlane, D.: Industrial adoption of agent-based technologies. IEEE Intelligent Systems 20(1), 27–35 (2005)
7. McCarthy, J.: What is artificial intelligence (2007), `http://www-formal.stanford.edu/jmc/whatisai/whatisai.html`
8. Monostori, L., Váncza, J., Kumara, S.: Agent-based systems for manufacturing. CIRP Annals - Manufacturing Technology 55(2), 697–720 (2006)
9. Schild, K., Bussmann, S.: Self-organization in manufacturing operations. Communications of the ACM 50(12), 74–79 (2007)
10. Wooldridge, M.: An Introduction to Multi-Agent Systems. John Wiley & Sons, Inc., New York (2002)

Chapter 14
Voice Conservation: Towards Creating a Speech-Aid System for Total Laryngectomees

Zdeněk Hanzlíček, Jan Romportl, and Jindřich Matoušek

Abstract. This paper describes the initial experiments on voice conservation of patients with laryngeal cancer in an advanced stage. The final aim is to create a speech-aid device which is able to "speak" with their former voices. Our initial work is focused on applicability of speech data from patients with an impaired vocal tract for the purposes of speech synthesis. Preliminary results indicate that appropriately selected synthesis method can successfully learn a new voice, even from speech data which is of a lower quality.

14.1 Introduction

Speech is a fundamental mean of human communication. However, healthy people usually do not fully appreciate how significant is the ability to speak with own natural voice for the social contact and interaction.

The human vocal tract is a complex and vulnerable system. Its damage can cause various problems with speech production. The damage can be related to an injury or it can be also an inevitable consequence of the treatment of another serious disease, such as mouth or neck cancer.

An example of such a radical treatment is the laryngectomy surgery. This medical intervention is performed on patients with laryngeal cancer when other less invasive types of treatments (e.g. radiation or chemotherapy) fail or are not possible. According to the extent of the carcinoma, various sections of larynx are necessary to

Zdeněk Hanzlíček · Jan Romportl · Jindřich Matoušek
Department of Cybernetics, Faculty of Applied Sciences, University of West Bohemia, Univerzitní 8, 306 14 Plzeň, Czech Republic
e-mail: {zhanzlic,jmatouse}@kky.zcu.cz

Jan Romportl
Department of Interdisciplinary Activities, New Technologies Research Centre, University of West Bohemia
e-mail: rompi@ntc.zcu.cz

J. Kelemen et al. (Eds.): Beyond Artificial Intelligence, TIEI 4, pp. 203–212.
DOI: 10.1007/978-3-642-34422-0_14 © Springer-Verlag Berlin Heidelberg 2013

be removed. In the case of so-called total laryngectomy, the removal of the whole larynx is performed. Patients who underwent this surgery are called laryngectomees.

An important part of larynx is the epiglottis – a valve which separates the respiratory and digestive tract during deglutition (swallowing). It protects against breathing in of swallowed food or fluid. This function of epiglottis cannot be reliably substituted. Thus, a permanent surgical separation of both tracts is performed during the laryngectomy surgery. After that, the laryngectomee breathes through a stoma – an opening in the trachea.

Another significant consequence of this surgery is the inability to produce speech in the common manner. It has two main reasons:

- Vocal folds were removed together with the larynx during laryngectomy. The vibrations of vocal folds during the expiration form an excitation component of speech that is then amplified and articulated into speech in the following parts of the vocal tract (primarily the oral cavity).
- Due to the detachment of the respiratory tract, the air from lungs cannot be passed through the mouth.

However, laryngectomees can learn and use several alternative ways of producing speech sounds:

- For producing **tracheo-esophageal speech**, a special voice prosthesis need to be placed between trachea and esophagus that were surgicaly separated during the laryngectomy. This prosthesis contains a one-way valve that allows the air to flow from lungs into the oral cavity. When the patient wants to talk, the tracheostoma has to be plugged and the exhaled air is pushed through the valve into the mouth where it is articulated into speech. The friction of air passed through the valve produces some vibrations and simulates the function of vocal folds. In principle, this method is similar to the natural way of speech production. The prosthesis needs to be regularly maintained and cleaned.
- **Esophageal speech** is produced solely by the upper part of digestive tract. First, the air is swallowed into the esophagus. Then, it is pushed back into the oral cavity for articulation. The produced speech resembles the belching. This method is very exacting due to the low capacity of esophagus.
- During production of **electrolaryngeal speech**, the function of vocal folds is substituted by an external device (*electrolarynx*) which is put to the neck where it produces intense mechanical vibrations. These vibrations are transmitted into the oral cavity while the speaker articulates.

All the aforementioned kinds of speech (also called *alaryngeal speech*) suffers from lack of naturalness and intelligibility. Although, the voice of two laryngectomees can slightly differ according to the various parameters (e.g. proportions) of the vocal tract, the new voice does not bear the former speaker identity. An important speech feature for distinguishing of speaker identity is the fundamental frequency f_0, i.e. the frequency of vocal folds oscilation.

Moreover, the changes and dynamics of f_0 are the essential means for expressing the prosody of an utterance. Thus, an important consequence of removing vocal folds is the inability to express the prosody. Neither electrolarynx nor various voice prosthesis are fully-fledged substitutions of vocal folds, because the frequency of electrolarynx vibrations is constant[1] and the air friction in the valve of the voice prosthesis is hard to control.

Recently, a new important task – alaryngeal speech enhancement – is solved in the field of computer speech processing [2, 6]. The objective is to improve the overall sound quality of the alaryngeal speech. The final aim is to create a speech-aid device which allows the laryngectomee to communicate with a more natural voice. This include recovering the speaker identity and adding adequate prosodic features.

14.2 Practical Aspects

The design of a speech-aid device for laryngectomees can be based on two different approaches:

- The device can detect/record the speech produced by the laryngectomee and the methods for its enhancing are based on the analysis and processing of that alaryngeal speech.
- The articulatory movements of mouth, tongue and other important parts of the vocal tract can be measured and corresponding speech reconstructed from those movements. The articulatory movements could be detected e.g. by using a suitably placed camera or by some specialized position and movement sensors attached to the crucial parts of the vocal tract. A great advantage of such an approach is that no alaryngeal speech need to be (loudly) produced. However, the problem of speech reconstruction from articulatory movements and firstly their sufficiently precise detection[2] is very complicated. Anyway, it could be employed in combination with the first approach as an additional information to the alaryngeal speech.

Hereinafter, we will deal solely with the first approach, i.e. creating the enhanced speech by processing of alaryngeal speech. From the practical point of view, the source alaryngeal speech should not disturb and coincide with the final enhanced speech. In the case of common types of alaryngeal speech, which is loud, this could be achieved by rotating stages of alaryngeal and enhanced speech. The control of switching the alaryngeal speech would be probably done directly by the laryngectomee. This simple approach has one significant advantage: If the switching is done phrase-by-phrase, a proper prosody of the utterance will be easier to estimate. However, it has also several relevant disadvantages:

[1] Recently, a modified electrolarynx device was introduced [7] which controls the frequency of vibrations by utilizing an air-pressure sensor measuring the air breathed out from tracheostoma.

[2] Moreover, the way of articulatory movements detection should not annoy the laryngectomee user of the device.

- Loud alaryngeal speech is still present and disturbs the communication.
- In the case of complex utterances, the incessant interruption and repetition of particular phrases could be confusing for the listeners (and maybe also for the speaker).

Those disadvantages can be surmounted by using so-called body conducted speech [1]. It is a special kind of electrolaryngeal speech where the standard electrolarynx excitation is substituted by a source of small-power vibrations. The speech is created by vibrations of air in the vocal tract passed on the soft tissues of the head or neck. It is nearly inaudible and cannot be used for inter-personal communication. However, it can be detect by special sensors, e.g. the NAM (non-audible murmur) microphone, which are attached to the body (usually to the head or neck) and measure the vibrations. In the case of body conducted speech, it would not be necessary to switch alaryngeal and enhanced speech because they do not disturb each other.

Disregarding the type of alaryngeal speech, there are two basic approaches to enhance alaryngeal speech:

- Using automatic **speech recognition**. Produced speech is first recognized by a specially designed ASR (automatic speech recognition) system. A text content of the utterance is extracted this way. Then, the recognized text is synthesised by a new voice. The knowledge of the utterance content allows to add some higher speech properties, e.g. prosody. An important disadvantage of such a speech-aid system is a delayed response caused by speech recognition process. It is caused by two main reasons:
  - Most modern ASR systems employ a language model, i.e. the probability of various word combinations is taken into account during the recognition process. Wrongly uttered and recognized words can be corrected this way. Depending on the complexity of language model, this can cause the delay of several words.
  - For a proper estimation of the prosodic features, it is necessary to know the structure of the sentence, at least partially. Thus the content of the whole utterance (or its relatively independent parts, such as phrases) has to be recognized first.
- **Speech signal transformation** without speech recognition. Spectral characteristics obtained by an analysis of alaryngeal speech signal can be repaired by using transformation methods. Then, new speech is reconstructed from those converted characteristics. During this process, neither the text content nor structure of the utterance need to be extracted. Thus, only a smaller delay caused by the computation can be expected. On the other hand, some types of voice defects contained in alaryngeal speech probably remain in resulting enhanced speech even after the transformation. Moreover, it is also not possible to estimate and add the proper prosodic features when the utterance content and structure are unknown.

14.3 Voice Conservation

A natural requirement for any speech-aid system is to produce a voice which is similar to the former voice of particular patients. Although every improvement of alaryngeal speech would be welcome, it is more motivating for the laryngectomees when they can identify themselfs with their new voices.

Most speech synthesis methods are able to produce voice with required speaker identity. Actually, it is a fundamental characteristic of those methods to learn and imitate the voice from the training speech data.

However, a quite huge amount of training data is necessary to learn the demanded voice. In the case of a laryngectomee, it could be a substantial problem to obtain a sufficient amount of speech data.

Naturally, speech recorded by the healthy voice (before the disease has broken out and harmed the voice quality) would be preferred. Some people could have recordings related to their job, e.g. various speeches, performances, presentations, or also some personal recordings, e.g. family events, reading fairy tales to their childern, etc. Unfortunately, the acoustic conditions of such audio data are often not optimal, which is not suitable for the purposes of speech synthesis – they contain a lot of background noise, the mutual position of the speaker and microphone is changing and the speaking style can be too expressive because it is naturally influenced by the ongoing event. Most people probably do not have any usable recordings at all.

Another solution is to record the speech data from patients after they are diagnosed with the laryngeal cancer. However, in the advanced stages of the disease, the vocal tract is usually significantly damaged and patients have usually serious problems with the speech production. The overall voice quality is poor or unstable and the speaking could be very exhausting for the patients. The voice could be also affected by the psychological condition of patients; they are often significantly stressed by their diagnosis and expected surgery. They could have serious troubles to concentrate on the recording process.

In early stages of the disease, patients are usually able to speak relatively well. In those stages, other less radical types of treatment (e.g. radiation or chemotherapy) are preferred; a laryngectomy is usually the last alternative. Recording those patients, who have a good prospect to be cured, could be counter-productive. They could be frightened because they could have a feeling that their health condition is worse and that they are candidates for the laryngectomy. That could negatively affect their mental state and complicate the progress of their treatment. On the other hand, this might be very individual and a significant number of patients may exhibit positive or at least neutral attitude towards their voice being recorded "just to be sure".

The last chance is to acquire recordings from another willing person with a similar voice but here a question arises, whether the patient will eventually be able to comfortably use someone else's voice

In any case, the problem of the convenient and considerate communication with patients should be discussed with psychiatrists and psychologists.

The process of obtaining and storing speech data of the patients can be called *voice conservation*. Yet it is just one from the series of fundamental steps essential for solving the complex problem of developing a speech-aid device for laryngectomees.

14.4 Proposed System Layout

In spite of all the problems, we have to settle for a practical solution that has a potential to be succesfully implemented and start helping the laryngectomees, perhaps not in an absolutely perfect way, but with a strong prospect of further significant improvements towards full recovery of the laryngectomees' ability to naturally and confidently communicate again using their own voice, without any psychological or social limitations.

We have to develop two somehow separable threads:

1. Framework for routine voice conservation and patient-oriented speech synthesis personalisation — in other words: how to effectively record and store the patient's voice (respecting all the clinical limitations) and create a speech synthesiser naturally speaking with this voice.
2. Actual speech prosthesis system development — how to integrate and innovate the existing speech technologies and push forward the state-of-the-art in this area.

As for (1), we have started cooperation with the Department of Otorhinolaryngology and Head and Neck Surgery, University Hospital Motol, Prague, and Department of Psychiatry, University Hospital Pilsen. These departments help us work with the patients under clinical practice, conduct recording experiments and establish facilities for patient acquisition and recording sessions. Our first results with speech synthesis personalisation are presented in the following sections of this paper.

Having a personalised speech synthesiser for a particular patient, the outline of the prosthetic device architecture (2) is following:

- The patient uses a source of small-power vibrations instead of the standard electrolarynx and articulates non-audible body conducted speech in a standard way using his articulators.
- Body conducted speech produced this way is detected e.g. by NAM and then with a slight delay transformed into its textual form by an ASR system.
- The textual form is immediately converted into naturally sounding speech by a text-to-speech (TTS) system using the patient's personalised speech synthesiser.

It is clear that the crucial role within the whole system is played by ASR and TTS systems. ASR must be able to reliably and in real time recognise body conducted speech produced by the patient, and TTS must be able to generate naturally sounding speech in the patient's voice, including estimation of appropriate prosodic features such as intonation (which cannot be controlled by the patient).

We have achieved very promising results in automatic recognition of audible electrolaryngeal speech [8], and therefore we have good reason to believe that the same ASR system can be adapted to non-audible body conducted electrolaryngeal speech, which is the focus of our current research. Moreover, since the user of a particular instance of the ASR system is always known, the speech recognition process can be strongly speaker-dependent and thus adapted to the particular speaker, which significantly decreases the error rate in comparison with speaker-independent recognition.

Functioning of the TTS system during the actual communication process is indeed far more robust and reliable than ASR. The core of the problems here lies rather in the system personalisation and the ability to generate highly natural speech in spite of generally low-quality speech data acquired from the patient.

14.5 Speech Synthesis

Modern speech synthesis methods (e.g. unit selection or statistical parametric speech synthesis [10]) need a large amount of training data (several hours of speech) to create a new system producing the desired voice in high quality. The standard recording procedure is demanding even for a healthy professional speaker because high-quality data are required for the purposes of speech synthesis. Recording of each utterance must be repeated until it is perfectly pronounced. Both the phonetics content and the required speaking style have to be kept. The overall recording process lasts for several weeks; a daily session takes about 4–5 hours.

It would be nearly impossible to record such a huge amount of speech data by a patient in an advanced stages of the laryngeal cancer. It would be too exhausting and they cannot also spend so much time with this. However, there are several alternatives with lower demands on the amount and quality of training data. The most promising one is using adaptation methods [9] within the statistical parametric speech synthesis [10]. This synthesis method (also known as HMM-based speech synthesis) employs statistical models (so-called hidden Markov models, HMMs) to represent the acoustic and other features of speech. These models are trained to match with the voice of given speech training data. Then, during the speech synthesis, trained models are employed to generate the required utterance.

Model adaptation methods enables to change the models from one voice to another. It can be performed by using significantly less speech data than would be needed for training of completely new models. First, source models are trained from speech data of a professional speaker, or data from more speakers can be used to trained so-called average models. Then, a transformation function (or a set of functions), that describes the main diferrencies between the original and new voice, is estimated from adaptation data. A great advantage of the adaptation is its lower sensitivity to the quality of the training speech data.

So basically we can record as much adaptation data from the patient as he or she is able to produce and then use them to modify the original models trained on a huge speech corpus of a professional speaker so as to achieve new models as closely matching the patient's voice as possible, yet still covering the whole extent of the original speech corpus.

14.6 Preliminary Experiments

For our first experiments, we apply experience from our previous work in the domain of statistical parametric speech synthesis [3, 4].

In cooperation with the University Hospital Motol, one voluntary female patient with laryngeal cancer diagnosis was selected and her speech recorded. Although methods for adaptation of statistical models are robust and quite resistant to imperfection of training data, quality of the data is still very important. A proper selection of recording place with good acoustical conditions is significant, therefore it was performed in an anechoic chamber for acoustical measuring and experiments.[3] Of course, it is not necessary to use a place with such perfect acoustical condition. A recording studio would be obviously more suitable because it usually provides more comfort to the recorded person, but for future routine operation of the whole voice conservation procedure we will have to settle for less adequate facilities available in the hospital environment.

The recording process had to be adjusted to the specific state of the patient. There are compromises that has to be done against the optimal setup, otherwise it would be impossible to record required amount of speech during one session. Consequently, the recorded utterances contain various stumbles, unexpected pauses and voice failures. About 500 utterances (approx. 1 hour of speech) were recorded during this session.

As it was already mentioned, in the future, we intend to record the patients in a suitable quiet room directly in the hospital. The main advantage is a better accessibility for the patients because they will not need to travel to another distant place. Moreover, they could record their utterances during several shorter sessions, which should guarantee a better quality of the speech data because the patients would not exert themselves to record the utterances all at once.

Besides lower speech quality caused by voice failures, recorded utterances contain a lot of significant reading or pronunciation errors. By a detailed analysis of the recorded utterances, we found out that more than 25 % were not read fully correctly. Due to this, at least a brief inspection of the recorded speech data is necessary to correct the transcription of such utterances or remove them from the training set before further processing. Several basic types of reading errors were recognized:

- 3.4 % utterances were unexpectedly interrupted by sudden long pauses. Probably the speaker lost the reading focus or was confused by an unexpected or difficult word combination.
- 7.4 % utterances were misspelled, i.e. contained at least one word that was pronounced with wrong or swapped phones. A common error was also the swapping of two consecutive syllables within one longer word.
- 1.4 % utterances contained a stumble inside a word. As a result, the word was interrupted by a pause or the corresponding affected phone is somehow distorted (e.g. unnaturally extended).

[3] The same place is usually used for recording of a professional healthy speaker – the whole process is in detail decribed in [5]).

- 5.7 % utterances contained a stumbled or misspelled word followed by its correction. Moreover, the repeated word was commonly unnaturally accented.
- 7.5 % sentences were not uttered in the required neutral style. In some cases, the speaker was influenced by the sentence content and spoke too expressively. In other cases, the speaking style was entirely unconcerned disrespecting the sentence structure.

Although HMM-based speech synthesis method is supposed to be robust and resistant to slight inaccuracies in training data, often or rough errors can significantly affect the accuracy of trained statistical models and also the quality of resulting synthetic speech. Some errors, that are related to the utterance content and do not affect the consistency of the speaking style (e.g. misspellings or unexpected pauses), can be remedied by the corresponding modification of the utterance transcription.

However, an appreciable portion of utterances (about 10 % according to our analysis) cannot be corrected this way. In the case of recording of a standard speech corpus by a professional speaker, those sentences would be recorded again. However, in the case of pre-laryngectomee patients, a prolongation or an additional repetition of the recording process would be frustrating. Therefore, we should expect that a certain part of recorded data would be unsuitable for the puposes of speech synthesis. The amout of unusable data can significantly vary for particular patients; it will depend on the extent of their disease, the related psychological condition and also their reading skills.

Within our first experiments we used about 30 minutes of speech to adapt models trained from 5 hours of speech from a professional female speaker. The synthetic speech produced by using the adapted models was definitely identified as the patient who took part in the experiment. Considering the utilised data, the quality was also acceptable. Moreover, the voice did not contain any noticeable voice failures or similar artifact caused by the vocal tract disease.

14.7 Conclusion

Our first experiments are promising. Voice conservation for patients with laryngeal cancer diagnosis, even when their speech is of lower quality, opens the possibility to create a speech-aid device producing the original personal voice that would otherwise be lost forever. Although the way to develop such a device is still long and a lot of research work has to be done, the current results can be already practically utilised. Laryngectomees can run the speech synthesizer with their own voice on their computers. This could be helpful in the postsurgical stage when the possibilities of inter-personal communication are very limited, which is indeed very frustrating. Moreover, the patients know that their voice is saved for future and they can have very justified hope that the technology will allow them to use it again.

References

1. Denby, B., Schultz, T., Honda, K., Hueber, T., Gilbert, J., Brumberg, J.: Silent speech interfaces. Speech Communication 52, 270–287 (2010)
2. Doi, H., Nakamura, K., Toda, T., Saruwatari, H., Shikano, K.: An Evaluation of Alaryngeal Speech Enhancement Methods based on Voice Conversion Techniques. In: Proceedings of ICASSP 2011, pp. 5136–5139 (2011)
3. Hanzlíček, Z.: Czech HMM-Based Speech Synthesis. In: Sojka, P., Horák, A., Kopeček, I., Pala, K. (eds.) TSD 2010. LNCS, vol. 6231, pp. 291–298. Springer, Heidelberg (2010)
4. Hanzlíček, Z.: Czech HMM-Based Speech Synthesis: Experiments with Model Adaptation. In: Habernal, I., Matoušek, V. (eds.) TSD 2011. LNCS, vol. 6836, pp. 107–114. Springer, Heidelberg (2011)
5. Matoušek, J., Romportl, J.: Recording and Annotation of Speech Corpus for Czech Unit Selection Speech Synthesis. In: Matoušek, V., Mautner, P. (eds.) TSD 2007. LNCS (LNAI), vol. 4629, pp. 326–333. Springer, Heidelberg (2007)
6. Nakamura, K., Toda, T., Saruwatari, H., Shikano, K.: Speaking Aid System for Total Laryngectomees using Voice Conversion of Body Transmitted Artificial Speech. In: Proceedings of Interspeech 2006, pp. 1395–1398 (2006)
7. Nakamura, K., Toda, T., Saruwatari, H., Shikano, K.: The use of air-pressure sensor in electrolaryngeal speech enhancement based on statistical voice conversion. In: Proceedings of Interspeech 2010, pp. 1628–1631 (2010)
8. Stanislav, P., Psutka, J.: Influence of different phoneme mappings on the recognition accuracy of electrolaryngeal speech. In: Proceedings of Sigmap 2012 (2012)
9. Yamagishi, J., Kobayashi, T., Nakano, Y., Ogata, K., Isogai, J.: Analysis of Speaker Adaptation Algorithms for HMM-Based Speech Synthesis and a Constrained SMAPLR Adaptation Algorithm. IEEE Transactions on Audio, Speech, and Language Processing 17, 66–83 (2009)
10. Zen, H., Tokuda, K., Black, A.W.: Review: Statistical parametric speech synthesis. Speech Communication 51, 1039–1064 (2009)

Chapter 15
Extended Mind: Is There Anything at All to Be Externalised?

Eva Zackova and Jan Romportl

Abstract. The paper discusses Clark's conception of extended mind and critically analyses his four criterions of externalised cognitive functions. Language as one of the most important means of externalisation is presented on the basis of Engelbart's conception of human enhancement. Clark's view of human–technology coupling is also strongly related to language which is regarded as a form of mind-transforming cognitive scaffolding. The material nature of language (as stressed by Clark in the domain of bodily gestures) is crucial for expression of mental concepts. This supports the belief that human cognitive enhancement is possible via technical means (e.g. AI-based speech prosthesis). However, from the philosophical point of view, even the categories of externality and internality of cognitive processes are very tricky.

15.1 Introduction

In computer science we can trace prognoses of man-machine symbiosis back to the 1960s when conceptions of so-called intelligence amplification occurred [1, 2]. They intended to reach the man and computer technology mergence in order to

Eva Zackova
Department of Interdisciplinary Activities, New Technologies Research Centre,
University of West Bohemia
Department of Philosophy, Faculty of Philosophy and Arts, University of West Bohemia
Univerzitní 8, 306 14 Plzeň, Czech Republic
e-mail: zacka@ntc.zcu.cz

Jan Romportl
Department of Interdisciplinary Activities, New Technologies Research Centre,
University of West Bohemia
Department of Cybernetics, Faculty of Applied Sciences, University of West Bohemia
Univerzitní 8, 306 14 Plzeň, Czech Republic
e-mail: rompi@ntc.zcu.cz

J. Kelemen et al. (Eds.): Beyond Artificial Intelligence, TIEI 4, pp. 213–221.
DOI: 10.1007/978-3-642-34422-0_15 © Springer-Verlag Berlin Heidelberg 2013

support and enhance current cognitive functions of human organism while further preserve actual human experience. Thereby the influence of the means used by human to facilitate pragmatic and epistemic actions (through artefacts and especially language) has been emphasised and opened to further discussion in the field of theory of mind.

The idea of mind being tightly coupled with its environment has already been—at least since the Maturana and Varela's autopoietic theory [3]—perceived and acclaimed as a sound and plausible alternative to cognitivistic theories of mind. The argumentation in favour of the extended mind [4, 5] is compatible with this paradigm of understanding of mind, and we believe that it is very important in the discussion about possibilities and frontiers of human enhancement.

In addition to theoretical analysis of the extended mind model, this chapter would like to illustrate our claims on an example of an AI-based system whose aim is to be able to adequately and in real time generate prosody for speech of patients after laryngectomy, who lost their ability to generate speech without external devices and to express their own prosody at all.

However, after having presented our arguments in favour of externalisation, we will have to ask an important question: is there anything at all to be externalised? Or is the dichotomy *internal–external* just an illusion?

15.2 Extended Mind Conception

We have focused on the conception of extended mind, which was firstly published in Clark's and Chalmers' essay *Extended Mind* (1998) and later more deeply elaborated by Clark in *Supersizing the Mind* (2008). The idea of extended mind is frequently compared with the embodied perspective [8], which Clark calls brainbound model because from this point of view all human cognition depends directly on neural activity alone, and mind processes are reduced just to a brain or neural activity [5].

Clark based his extended model on so-called parity principle, which held that "if a process in the world works in a way that we should count as a cognitive process if it were done in the head, then we should count it as a cognitive process all the same" [5]. In brief, "if a certain state plays the same causal role in the cognitive network as a mental state, then there is a presumption of mentality" [5]. Hence we can infer a thesis that not all our thoughts, not all our cognition have to be located in our head.

15.3 Express Yourself—Exted Yourself: Language as a Mind-Transformer

At least since the mid 20th century speech competence is regarded as probably the most complex cognitive ability of human brain. In that time, an extremely influential Whorfian principle of linguistic relativism and determinism popular mainly in the field of linguistic disciplines and social sciences was frequently adopted by computer scientists and AI researchers as well. For example, Douglas Engelbart devoted a great portion of attention to the role of both human and computer language processing in his work on human enhancement and intelligence amplification through technology [1].

Engelbart believes, in accordance with the Whorfian hypothesis, that language determines our world-view and our thinking, but in addition, he stresses out (apparently inspired by Korzybski) the crucial role of intellectual capability to manipulate symbols [1]. From this point of view, it is taken for granted that the biggest progress in cognitive processes / human cognition was done in the moment when language emerged in its written (materialized) form and thus *external* symbol manipulation was enabled.

In the next step, Engelbart suggests a deliberate integration of digital computer into this process for the purpose of making the external symbol manipulation automatic. Of course, this has become a commonplace for us nowadays; we consider evident that an immense advantage has been taken from automatic computer language processing and symbol manipulation in general, which proves Engelbart's original vision of new and more effective language forms created by/for computers to solve and facilitate various problems and tasks too demanding or even unsolvable by a human.

In fact, the way the computer is integrated into human's action is much deeper in Engelbart's project of man–computer symbiosis. He designed a so called H-LAM/T system where Human uses Language, Artefacts and Methodology, which are further developed by Training to augment human capabilities and cope with the world. Although the computer operates in this system primarily as an artefact similar (but radically more sophisticated) to a pencil and piece of paper, regarding the language it also provides its own artificial forms of languages to represent the world.

The whole system is described by Engelbart as synergistic and working in a feedback loop. Any new way of material manipulation of symbols (a new cognitive action) constantly affects language and thinking itself and brings them to higher and more effective levels. For example, we have recently globally adopted a new word to refer to a specific cognitive action focused on effective information searching that we learned to use by force of using computers. Thus the verb to *google* is now a part of our vocabulary and influences the structure of our reasoning and thinking about the world, which will influence our language again in the future.

Marshall McLuhan was definitely on the same page in his book *Understanding Media: The Extensions of Man* from 1964. His famous slogan *medium is a message*, when medium is meant as any technology, and any technology (especially communication technology) as an extension of man, illustrates the general spirit of that time also very well [6].

These ideas of essentially natural coupling of human and technology, ideas of "natural-born cyborgs" [5, p. 42], which arose fifty years ago, are still attractive and enthusiastically advocated by many. Within the Clark's conception of extended mind, they are further elaborated with respect to present neuro- and cogno- sciences and philosophy of mind. Clark approaches the problem of making people better in their physical and mental performance (human enhancement) with the aid of technology from the transhumanistic perspective. He rejects any fixed standard of our embodied existence which would be merely supported by various tools. Instead, he believes "... that human minds and bodies are essentially open to episodes of deep and transformative restructuring in which new equipment (both physical and 'mental') can become quite literally incorporated into thinking and acting systems that we identify as our minds and bodies ..." [5, pp. 30–31].

Once this potentiality of human agent and equipment mergence is realized, the new "systemic whole" is created [5, pp. 33–35]. Clark places one requirement on the equipment. It has to be *transparent*, in the sense of "without (generally) being itself an object of conscious thought or effortful control" of an agent [5, p. 33]. Then we can talk about true incorporation of the equipment, which allows building a new "agent–world circuit" enhancing our capabilities by bringing the interface from human–world line to systemic whole–world line [5, pp. 31–32]. As we interact with the world through this new interface, we have to adapt our mind and body to the new experience—we have to *negotiate* our body through a chain of feedbacks, which is, according to Clark, quite easy and natural for us thanks to brain plasticity and general "cognitive permeability" [5, p. 40].

Apart from similarities of Clark's and Engelbart's work in systemic approach to human body and mind, in emphasizing deep coupling between man and technology resulting in profound restructuring of thinking and acting, we can trace the same strong interest in human speech competence in Clark's conception of extended mind.

Clark proposes to consider language as a "form of mind-transforming cognitive scaffolding" [5, p. 44] that enables us to label the world (in Engelbart's words: to create symbols of our non-verbal pre-concepts, to generalize and to manipulate symbols externally), to develop expertise and to control our own thinking process [5, p. 44], and that helps "provide the tools we use to discover and build the myriad other props and scaffolding" [5, p. 60]. This is pretty much in compliance with linguistic determinism we have mentioned above. Furthermore, Clark tends to favour the thesis of Fodor's *mind modularity* and the indispensable role of language as a tool for integration of information from multiple knowledge bases [5, p. 49]. The key feature of language is again its material character, which makes both the extension and the externalisation of human mind possible.

Besides the aforementioned advantages of labelling, structuring, manipulating and integrating role of language in cognition, Clark concerns "the role of bodily gesture in thought and reason" [5, p. 123]. This look into the (in fact) paralinguistic dimension of language could be quite challenging for our own research ambition to develop speech prosthetic device with automatic production of prosody of speech.

In his promotion of gesturing as a crucial part of thinking itself, Clark refers to a number of neuroscientific studies which reveal that gestures are not a mere additional form of expression to the verbal form of language, just supporting the communicative situation between two participants. He supports the claim that by gesturing we actually pre-realize and shape our own thought even before we become conscious of it in a process of conceptualizing and verbalizing the thought. According to Clark we can further infer that bodily gesture is a true instance of a cognition progressing externally out of our head.

Another supportive example for extended mind conception related to a speech competence can be found in Gazzaniga's experiments with epileptic patients. These experiments, when sectioning of corpus callosum (connecting the left and the right cerebral hemispheres) is performed in order to stop epileptic seizures, suggest that the transferring role of corpus callosum, which is necessary to complete a basic cognitive action of recognition of visually perceived object and its subsequent verbal labelling, can be preserved by its externalisation. In this particular case, a part of cognitive process is literally done on a piece of paper (for details see [7]).

All the arguments in favour of truly extensional character of our mind are heading to a claim that the mind, the body and the world have been coupling in various ways from everlasting and, in fact, they are supposed to serve the idea of fully functional and intimate two-way wiring of a man's brain to a technical device realizing kind of a cognitive process. Together with Clark, we believe this is possible, indeed. Furthermore, concerning the paralinguistics as a pivotal part of language, and language as epitomisation of our mind's potency to engage the outside physical world in developing the mind's own inherent processes and directing them to a particular goal in a feedback loop, we believe that the externalization of prosody production can be regarded as another convincing argument for extended mind conception as well.

15.4 Criterions of External Cognitive Process

Clark defines four basic criterions of what he still counts as a cognitive process even though it is realized outside our brain. In our research we have slightly redefined them and suggested our own criterion for what can be considered as an external cognitive process—we called it "hard-wired system criterion".

The first Clark's criterion says that the resource must be reliably available and typically invoked [5]. The criterion of typical invocation means that one would never say "I don't know, I have to check my external memory" when one needs to consult an external resource to get the answer. Instead, the resource is automatically invoked, and only in the case that the answer is not found, then one would say that s/he does not know—just like in case of biological memory.

The second criterion says that "any information thus retrieved must be automatically endorsed and trustworthy as something retrieved clearly from biological memory" [5]. Automatic endorsement criterion seems reasonable to us, even though we believe that the external memory can be even more trustworthy than the biological one. However, the key issue, which is commonly ignored, is that this criterion is related just to a *memory function*. From our point of view, this is a serious deficiency of the Clark's criterion. Obviously, we know other types of cognitive processes which are more complex than retrieving information from our memory—as for example in case of speech competence, attention, recognition, learning, reasoning and so on.

The third criterion says that "the information contained in the resource should be easily accessible as and when required" [5]. We believe that this one is plausible enough, it is quite close to the first criterion and we have accepted them both.

The last criterion claims that "the information must have been consciously endorsed at some point in the past and be there as a consequence of this endorsement" [5]. We strongly disagree with this condition. According to us, the information does not have to be consciously endorsed every time. Firstly, a general endorsement can be done instead of a recurrent particular endorsement. For instance, we can decide to trust all information from a particular source. Hence we do not have to make the endorsement for each particular piece of information or process. Secondly, and more importantly, we hold that we can externalise the act of endorsement itself (for example in cases of attention disorders or in case of certain levels of externalised speech competence).

15.5 Hard-Wired System Criterion

Since we regard Clark's criterions as somehow slightly missing our view of the extended mind conception, we have suggested a criterion of a "hard-wired system" (or simply "hard-wired system criterion"), which is still based on the aforementioned parity principle and Clark's original requirements that we consider as reasonable, such as easy accessibility, reliability and typical invocation of the resource, but which we have enriched with mutual adaptation (of the externalised system and of the mind processes) and functional dependency (of the brain cognitive processes on the external cognitive processes).

To summarise our idea, the hard-wired criterion claims that we can speak about extended mind when functioning of the external system (which is based on the parity principle) influences functioning of the brain and mind in such a way that other mechanisms of the brain adapt themselves so that in case the system is disconnected, it will cause serious deficiencies in cognitive abilities of the person.

This criterion thus presupposes a multilayered structure of feedbacks preserving "homeostasis" within the system consisting of the human coupled with the artificial device that is "hard-wired" (metaphorically) to him. These feedbacks are realised by both the human and the device. If the device is disconnected, the homeostasis is destroyed, at least until some other higher-level feedbacks manage to adapt what remains of the original system (i.e. the human plus some other possible devices) to the new situation. It is thus somehow questionable whether we can still regard the mind as being modular in terms of "plugging" and "unplugging" various modules—the cognitive functions are realized by the whole mind in the state of homeostasis, not by the particular modules, even though the "modules" take part in it.

15.6 Speech Prosthetics as a Real-Life Application of Extended Mind

To foster the extended mind conception defined by our hard-wired system criterion more deeply, we illustrate it by a concrete example of a real-life application.

We work on a system that will allow patients after total laryngectomy to use intelligent speech prosthetics based on TTS (text-to-speech), ASR (automatic speech recognition) and AI technologies (cf. Chap. 14 of the present volume).

The system will transform the distorted and hardly comprehensible speech of laryngectomees to naturally sounding synthesised speech. However, the patients will have absolutely no control over the output prosody, which will be completely estimated by the artificial system [9]. This means that the internal cognitive process, normally responsible for prosody generation, will no longer be used and will be substituted by the external cognitive process realized by the speech prosthetic system, creating tight man–machine coupling in the domain of speech competence.

Prosody as the whole set of suprasegmental features of speech (intonation, intensity, timing and others) is very interesting in that it partly belongs to the domain of language and partly to the domain of paralinguistics. It thus exhibits the same dispositions for externalisation in the sense of Engelbart, but it also has quite close relation to non-verbal features of human communication such as gestures discussed by Clark.

We believe that this is a good example of AI-based externalisation of a cognitive function of human mind and that it could be regarded not just as treatment but also as enhancement. Furthermore, this example involves a cognitive function that is more complex than in case of memory.

15.7 Internal and External

One can see that the whole argumentation in favour of externalisation can be quite succesfull. However, this steals attention from another very important question: if we argue that some cognitive processes can be external (hence opening way for extended mind), we first have to be convinced that there are some internal cognitive processes, that this internality is "normal" or "default", and most importantly that there exists a place to which we relate the concepts of "internal" and "external". If something goes on or happens inside this place, it is internal, otherwise it is external.

But what is this place with regard to our cognition? We simply cannot say that it is our head even though we feel that our consciousness is located there and that we observe the world from there. We cannot separate the head from its body (metaphorically as well as literally) without destroying all the cognitive processes.

So we can try it the other way round: what all can a person loose from his body and still maintain cognition? It can be hair, fingernails, perhaps fingers, maybe limbs, some organs, even the heart as long as surgeons are able to replace it by pumps. But we can also loose neurons, and we do—lots of them. Therefore, this way we still cannot delimit what is internal and what external. It is like with a musical band: all its member can change and still it is more or less the same band as long as it plays the same music, has the same fans and performs the same tours.

We simply want to say that this place, this delimitation is not the body. What goes on inside the body is not internal, especially with regards to cognition. Lets imagine a silicon chip with some kind of a neural interface performing particular cognitive functions, perhaps in the sense of enhancement. Does it make a difference whether this chip is placed under the skull, under the skin, or worn in a shirt pocket of the user? Indeed this cannot create the basis for determining the internal from the external. Yet the decision also cannot be based on the material which the device is made of—"artificial" (e.g. silicon chip) meaning external, "natural" (e.g. brain tissue) meaning internal. Although it is not possible (now, and perhaps never will be), we can at least imagine a situation when medical science is able to save and preserve the brain of a person whose skull was destroyed in an accident. Does this brain—placed in a jar connected to the body by cables—perform internal, or external cognitive processes?

Anyway, all these examples are still not very important because the core of the problem lies elsewhere: it is not *thoughts* that happen inside the head—only firing neurons, which is something very different from thoughts, in the same way as digestion is different from stomach. However, we still must acknowledge presence of some *phenomenological internality* in our consciousness. We simply know that something is inside our mind and something is not. And here comes the most important point: we have to speak about *mind*, not about *head*. The place for our cognitive internality is our *consciousness*—the essence of *self* that places *my* cognition *here* and *now*. Apart from this, there is no other role of this essence, this *eidos*.

This however means that everything else is external. Every neuron, no matter how deep in the head, no matter how much "brain-bound", is external to the essence of consciousness, every cognitive process outcome is external too. Indeed there would be no consciousness and no self without those neurons and cognitive processes. We only say that they all are and always have been external.

And so is there anything at all to be externalised? Philosophically, no, because everything is already external. Pragmatically and technically, yes, because it is a real challenge for new technologies and one should in the future care if he was born with the ability to see ten miles in absolute darkness, or this ability was given to him "externally".

References

1. Engelbart, D.C.: Augmenting Human Intellect: A Conceptional Framework. Prepared for: Director of Information Sciences Air force Office of Scientific Research, Washington (1962)
2. Licklider, J.C.R.: Man-Computer Symbiosis. In: Taylor, R.W. (ed.) In Memoriam: J.C.R. Licklider: 1915-1990. Digital Systems Research Center Reports 61, Palo Alto, CA (1990)
3. Varela, F.J., Thompson, E., Rosch, E.: The Embodied Mind. MIT Press, Cambridge (1991)
4. Clark, A., Chalmers, D.: The Extended Mind. Analysis 58(1), 7–19 (1998)
5. Clark, A.: Supersizing the Mind: Embodiment, Action, and Cognitive Extension. Oxford University Press, New York (2008)
6. McLuhan, M.: Understanding Media: The Extensions of Man, 1st edn. McGraw Hill, New York (1994); reissued by MIT Press

7. Gazzaniga, M.: The Social Brain. Basic Books, New York (1985)
8. Rupert, R.D.: Cognitive Systems and the Extended Mind. Oxford University Press, New York (2009)
9. Romportl, J.: On the Objectivity of Prosodic Phrases. In: The Phonetician, vol. 96, pp. 7–19. The International Society of Phonetic Sciences (2010)

Chapter 16
Embodied Agent or Master of Puppets: Human in Relation with his Avatar

Mateusz Woźniak

Abstract. Neuronal and psychological processing of mental states of human operators of avatars is a relatively new but fast growing topic of interest. Most of the research focus on avatars meant as virtual agents controlled by humans, but the same explanations should also apply to other types, such as human-controlled robots. The most recent neuroscientific research proves that processing of avatar in-formation tends to increase activation in brain areas responsible for processing in-formation about body enhancements and embellishments. The author presents results of his behavioral studies and argues that the effect of treating an avatar as a body enhancement can be mediated by cognitive content such as attitudes, convictions and beliefs, causing different prospective behavior.

16.1 Introduction

People engage in virtual interactions with artificial agents more and more frequently, and possibly with each passing year such interactions will be more and more common. Modern human has the possibility to encounter a human-like robot, but in most cases his interactions with AI are the ones he experiences when using a personal computer. Probably the most obvious example of such interactions are computer games, which allow a gamer to explore virtual worlds and meet digital people. Other examples can be found on the Internet. Chatterbots such as Alicebot [31] (ALICE - Artificial Linguistic Internet Computer Entity) or Jabberwacky [33] give an opportunity to participate in a text conversation with an "artificial intelligence". Similarly, when interacting with the so-called Embodied Conversational Agents we can not only speak to but also interact face-to-face with virtual "people" [4]. One example is GRETA [32]—a computer animated woman capable of keeping up a verbal conversation and (based on information from cameras) responding with facial

Mateusz Woźniak
Institute of Psychology, Jagiellonian University, Al. Mickiewicza 3, 31-120 Kraków, Poland
e-mail: mgwozniak@gmail.com

J. Kelemen et al. (Eds.): Beyond Artificial Intelligence, TIEI 4, pp. 223–235.
DOI: 10.1007/978-3-642-34422-0_16 © Springer-Verlag Berlin Heidelberg 2013

expressions and gestures to the body language of her human interlocutor. It is one of the last steps to a fully convincing face-to-face interaction with a virtual person.

In the examples described above, a human communicating with an artificial being enters the interaction using his physical identity, he is acting as himself. But it does not necessarily have to be the case. Sometimes we interact with other agents through an avatar [2]. Like a professional puppeteer, we can use various objects to act on our behalf. Then we interact face-to-face, but using the face of our avatar instead of our real one. Moreover, we can take control over a relatively "independent" virtual being—we do not have to treat an avatar controlled by us like our costume, but we can assume the position of a voodoo sorcerer who temporarily takes control over somebody's (although artificial!) life. This raises a question—what are the factors increasing the probability of each interpretation and, even more importantly, what are the possible consequences of each one? Within this text and in the research described, the emphasis was put on computer avatars, but the conclusions can be easily and effectively extrapolated onto other analogous situations, including taking control over a robot.

16.2 Treating Artificial as Real

Recent research has provided a great amount of evidence showing that humans tend to intuitively and naturally engage in interactions with artificial agents as if they were real (*Media Equation Theory* [21]). "Human interactions with computers, television and new media are basically as social and natural as interactions in normal life" [21] (p. 15); moreover, this is the case even though people engaging in such interactions explicitly state that they do not consider their interaction partners to be living or intelligent beings. The consequence is that people tend to apply many human traits to machines or virtual agents, in a process frequently referred to as anthropomorphism [28], which subsequently results in a behavior different from that which would be expected from a strictly task-oriented, depersonalized approach.

It is important to mention that, regarding artificial agents, anthropomorphization can be understood in two different contexts: communicative and visual. Conversational agents like Jabberwacky or ALICE are examples of the former. These programs neither have a body in any form nor do they possess any representation of a body. They create an illusion of being human by being able to use language. The greater their "linguistic skills", the bigger the possibility that they will be treated as human beings. This is also the basic idea behind Turing's test [27].

On the other hand, the term "anthropomorphism" is actually more frequently investigated as a potential of visual and spatial stimuli to be perceived as a living object, preferably a human entity. For example, research conducted by Morewedge, Preston and Wegner [20] shows that an important factor of humanization is the physical similarity to a human and appropriate movement characteristics of an agent. These insights are further strengthened by neuroimaging studies. Gazzola et al. [12] have conducted an fMRI study in which they were able to observe activation in the mirror neuron system in participants observing the movements of a robot when the dynamics of these movements were close to biological motion. Research indicates that the mirror system is probably responsible for, among others, such traits as empathy towards or imitation of an observed object. This line of research concentrates on the requirements which an image or object has to fulfill in order to be convincing as an actor.

Usually, humanoid artificial agents have to be convincing in both of these contexts—communicative and visual. Embodied Conversational Agents like GRETA (and most of the Non-Player Characters in computer games) aim to become as realistic as possible (or necessary) both visually and as conversation partners. On the other hand, other research shows that the requirements to generate an emotional reaction in humans in response to the behavior of virtual agents can be surprisingly low. For example, in a study conducted by Zadro, Williams and Richardson [30] participants took part in a very simple ball-tossing game with two other (virtual) players. During the game, the virtual players started ignoring the participants (they stopped passing the ball to them). As a result, the participants experienced negative affect and feelings of being ostracized, even though they knew they were not playing with real people.

16.3 Avatar—Virtual Body

Among artificial agents, a special set of entities can be distinguished—avatars, which are virtual agents controlled by us. Interactions with other people or subjects within digital (or artificial) realms seem to correspond to our everyday experiences, but in case of avatars controlled by us it is a little harder to find an analogy in the real world. It is not usual to "jump into" somebody else, into his body, identity and personality. Despite this, new technologies allow such bizarre situations; moreover, people experiencing them appear to feel surprisingly comfortable. After all, computer entertainment, especially computer role-playing games, makes it possible. Scientific investigation concerning the topic of avatars has begun only recently, but some initial results are already available. For example Yee and Bailenson [29] have introduced the term "Proteus effect". They have conducted an experiment in which participants entered a highly immersive virtual environment using virtual helmets. They then had some time to acquaint themselves with how their digital bodies look like and afterward they were given an experimental task to perform. They had to stand at a comfortable distance from another avatar and tell him as much about themselves as they felt comfortable with. As a result, participants controlling "attractive" avatars maintained, on average, shorter distance and revealed more information about themselves than their counterpart controlling "unattractive" avatars. According to Daryl Bem's Self-Perception Theory [3] they modified their behavior in order to fit their (even though artificial and temporary) look. They behaved as if they were "clothed" in another person's body.

The latest inquiries into the relation with virtual bodies focus on the neural underpinnings of this process. In a recent article Ganesh et al. [13] have analyzed fMRI data regarding brain responses from two groups: intensive World of Warcraft gamers and non-gamers. The first group (the gamers) was analyzed concerning their responses to WoW avatars, while the second group (the non-gamers) was analyzed concerning their responses to known and liked cartoon characters. The result was that WoW players responded with greater activation in the left angular gyrus of the inferior parietal lobe when rating their avatars than when rating themselves and other people they were close to. Furthermore, the magnitude of this avatar-referential activity was positively related to their level of body plasticity measured by a questionnaire estimating the ease to incorporate and self-identify with body enhancements (for example prostheses or tattoos). It suggests that, at least to some extent, avatars are perceived as extensions of the body.

16.4 Body Representations

Gallagher [10] has proposed and Synofzik, Newen and Vogeley [24, 25] have later developed the distinction between the so called Sense-of-Ownership (SoO) and Sense-of-Agency (SoA). This distinction was initially intended to help understand physical embodiment, including clinical conditions such as alien and anarchic hand syndrome [16], but seems also very useful when analyzing avatars and virtual reality. Basically, SoO is the feeling that, for example, "my hand belongs to me", while SoA is the feeling that "I am in control of my hand" (and not some other force, like another person's hand raising and lowering my arm). The contribution made by Synofzik et al. [24, 25] was a further distinction between the levels of "feeling" and "judgment" for both SoO and SoA. The alien hand syndrome is a clinical condition manifested by the feeling of estrangement from one's own hand. A person suffering from it has a very strong impression that his or her hand does not belong to him or her. Such a person *feels* as though the hand belonged to somebody else, but at the same time *knows* that it is, in fact, his or her hand. In this case the term Feeling-of-Ownership describes their impaired attribution on the lower, sensory level, while Judgment-of-Ownership represents their healthy cognitive judgment. Moreover, their Sense-of Agency is unimpaired. In case of anarchic hand syndrome, the Sense-of-Ownership is basically preserved on both levels. What is impaired is the mechanism behind the Sense-of-Agency. A person with an anarchic hand does not have control over one of his or her hands (usually the left one), and this hand seems to possess "a mind of it's own". It can be unzipping a jacked while a person tries to zip it with the healthy hand, or taking off the clothes when a person tries to put them on. In this case the Sense-of-Agency is impaired on both levels. Such a person neither has a low-level Feeling-of-Agency, nor the higher-cognitive-level Judgment-of-Agency, which simply means that he or she neither feels nor thinks him- or herself capable of controlling the anarchic hand.

When controlling an avatar a person should, at least to a considerable extent, experience Sense-of-Agency on both levels. The quality of feedback between the controller and the visual output determines how strong SoA becomes. Regarding the Sense-of-Ownership: a virtual body exists outside of the physical body, therefore the Feeling-of-Ownership cannot be physiologically felt. At first, it appears to be the case also with the Judgment-of-Ownership—determining what belongs to a body and what does not, ought to be fairly straightforward. A problem arises, however, when prosthesis, implants and transplants are taken into account; even the simplest case can become ambiguous. After a successful liver transplantation the new organ seems to become a part of the body and likely most patients, after a short period of time, interpret the situation accordingly and consider the "alien" liver a part of themselves. However, among the society of Kanaks living in New Caledonia described by Maurice Leenhardt [14] the liver had a very significant meaning. For them it would have probably been impossible to accept the new liver as a part of themselves on a cognitive level. Likewise, the judgment of whether a prosthesis or a tattoo has became a part of the body seems to depend mostly on the individual beliefs rather than being an arbitrary decision. And if a prosthesis can become a part of the body representation (on the level of Judgment-of-Ownership), than possibly an avatar can as well.

Before proceeding it will be useful to precise how a body can be represented in the human brain or mind. Gallagher [11] has distinguished between two types of body representations: body schema and body image. Body schema is a centre for sensory-motor information. It is a sensory-motor representation which guides the actions. When a person loses a leg and experiences a phantom pain, it is the body schema which "keeps" a representation of the lost limb (also similar to the concept of a neuromatrix by Melzack [18]). It is localized primarily in the sensory-motor cortex. The body image consists of all the remaining, "higher" representations. Others, like Schwoebel and Coslett [22] (or [8] for review), have improved this classification by dividing the body image into visuo-spatial body image and a semantic representation. The visuo-spatial body image is a structural representation of a body, mostly visual but also with somatic perception. It holds information about the relationship between the body parts and how a body looks like, making it possible to recognize oneself in a mirror. It is relatively stable, but changes during the course of a lifetime (we look differently as children, adults and elders). The exact brain localization of this structure (and whether it is a single localization) is unknown, but some research points at the significance of the Temporal-Parietal Junction region [5, 26]. The semantic representation holds all symbolic information regarding our bodies. It is conceptual and linguistic. In classical social psychology of self and identity these two representations (visuo-spatial and semantic) do not seem to be distinguished. Basing on the concept of self-schemata introduced by Marcus [17], Altabe and Thompson [1] understood the body image simply as a cognitive structure representing self-knowledge about one's body. This cognitive version was chosen when designing our own research described later.

Feeling-of-Ownership and Judgment-of-Ownership can be understood as individual phenomenological percepts (FoO) and beliefs (JoA) that certain object is a part of one of the body representations. Unfortunately, it becomes much more complicated when attempting to draw a line between tools, body extensions, and objects incorporated into a body (see [9] for brief overview). For example, recent research shows that using a tool causes temporary changes in the body schema [15]. Regarding the problem of whether an object can become incorporated into an existing body model or should be treated as its extension, De Preester and Tsakiris [7] propose (in conjunction with the existing research data from 2009) that incorporation can happen only for objects that replace the original body parts and fit with the pre-existing normative body representation. This way a prosthesis can become incorporated, but a tool cannot. A tool can become a temporary online [6] extension of a body, which is not limited by the norms and strict borders of a body-model. Using this reasoning, an avatar can only be seen as an extension of the body. On the other hand, as described previously, research conducted by Ganesh [13] shows that among gamers playing massive multiplayer online role-playing games (MMORPG) like World of Warcraft, thinking about an avatar can increase activation in brain region responsive to prosthesis and body embellishments which basically can be perceived as incorporated. Further research must be conducted in order to say whether an avatar can be treated as incorporated or just enhancement of a body and whether it can become a part of a stable offline representation.

16.5 Research

16.5.1 Purpose and Methods

In order to determine whether situational cues can play a role in establishing a relation between the person and an avatar an experimental study was conducted. The participants were told that they are going to play a computer game. A computer game modification was used in order to create a virtual variation on a classic obedience-to-authority experiment, originally conducted by Stanley Milgram [19] in the 1963. The participants' task was to, using a virtual avatar, teach another digital person (the learner) a set of pairs of words and then to test his knowledge of those pairs.

The computer program which was used consisted of three parts, each taking place in a different virtual room. The first room served as a tutorial to controls and interface. It contained a virtual experimenter and a computer. The virtual experimenter gave instructions on how to move around, initiate dialogue and use a computer. He also charged the participant with a task, which was to approach the computer, use it and perform a few simple arithmetic calculations (this task was set in order to initiate the participant into the game's world and to give the real experimenter time to make sure that the player understands the game's mechanics). Upon completion, the computer instructed the participant to report to the virtual experimenter. Following this instruction caused the program to switch to the second phase of the experiment.

In the second room, the virtual experimenter asked the participant to initiate dialogue with the "learner" and to read him a list of 75 pairs of words. The instruction was presented as follows:

> I am conducting a research concerning the influence of punishment on memory. In this experiment, you will be a teacher. Your first task will be to read to the person, who will be a learner in this experiment, a list of 75 pairs of words. This person is standing nearby. Here is the list (he gives you a piece of paper). Please tell me when you have read all the pairs.

Upon initiating dialogue with the learner, he introduced himself by saying "Good day. My name is Christopher and I am also taking part in this experiment.", to which the subject replied "Pleased to meet you. I'm <name>". <name> was, depending on the version of the experiment, either the participant's actual name or number 101115. Teaching the learner had the form of an in-game dialogue, during which all participants were informed that "it is not necessary to read the pairs of words. A click on the proper dialogue option suffices to symbolically teach them to the learner, since he will be asked about them in the next step". Upon completing the "teaching" process, the player was asked again to report to the virtual experimenter, and subsequently taken to the third and final room. Up until that point the real experimenter was present and ready to help clear up any confusion caused by controls or game's mechanics. At the beginning of the third phase, however, the subject was addressed with the following words: "From this point onward, I will leave you alone. In case you have any doubts or questions please address the virtual experimenter. Please call me only if asked to do so by the virtual experimenter". At this point, the real experimenter left the room.

The third phase of the experiment was a modified virtual version of Stanley Milgram's 1963 obedience to authority experiment [19]. The virtual experimenter began by explaining the task:

> I am conducting research concerning the influence of punishment on memory. In a nearby room there is a participant—a learner. He is connected to a machine capable of administering electric shocks. Your task is to ask him questions and react to his answers. If the correct answer is given, proceed to the next question. In case of an incorrect answer, you will first administer an electric shock and after that ask the next question. For the first incorrect answer you will administer a 30-volt shock. Any subsequent shock will be 30 volts stronger than the previous one, up to the strongest at 450 volts. If you are ready, please approach the computer.

The original experiment had 15-volt increments between shocks; however, after the pilot study it was concluded that a virtual version reflecting this condition would be too long. The participants asked the questions through the dialogue mode screen, after which a short animation of the learner either giving an answer or reacting to the shock was played. The learner's answers were displayed as a floating text above his head, white in color until a 150-volt shock was administered, and red afterwards. No sound effects were used at any point in the game. Anytime, except when the animations were played, the participant could, using the avatar, step away from the virtual computer and talk to the virtual experimenter. The participant could then ask if the learner can be hurt (the answer given was taken directly from the original experiment: "There is no permanent tissue damage."), inform the virtual experimenter that the learner does not wish to continue (to which he replied: "Whether he likes it or not, the learner must continue until he answers all the questions correctly") and refuse further participation. Unlike in the original experiment, quitting the experiment was not particularly hindered (because the goal was not to measure obedience to authority). Selecting the "I want to stop the experiment" option resulted in the virtual experimenter replying "You must complete the experiment" with two dialogue options below: "[I refuse to participate any further in the experiment]" and

"[Continue the experiment]". Choosing the former meant halting the experiment. An alternative possibility to complete the experiment was to administer the 450-volt shock.

Upon completing the experiment, the virtual experimenter asked the participant to call back the real experimenter, who gave them an ending questionnaire. It consisted of 9 questions, of which the first two were used to collect qualitative data (these questions were: "Did you administer the strongest shock and why?" and "What are your feelings concerning this experiment?"), while the remaining 7 were a typical multiple-choice questions. In addition, basic demographic data (gender, age, occupation and education) was collected. Once the questionnaire was filled in, the subjects were shown a short animation in which the "learner" informed them that he was not hurt and bode them good-bye. Afterwards, the participants were asked to give their impressions of the experiment. This was followed by an explanation of the experiment's purpose as well as a short lecture about Milgram's original experiment given by the experimenter. In case the participant asked for more detailed information, a prolonged conversation took place during which the experimenter attempted to address all arising questions and doubts.

There were two independent variables. First one was the name of an avatar. Two extreme situations were possible. In one condition a participant was introducing himself with his or her real name while in the second condition—with a fixed number. The intention was to differentiate between a situation of identification with an avatar (avatar becomes part of the self image understood as a cognitive representation of one's body) and situation of treating an avatar as a nameless puppet. Such manipulation was intended to prime one of these cognitive interpretations of relation with an avatar. The second variable was the outfit of an avatar—it could have been civilian or military. The purpose of including this variable was to additionally test the Proteus Effect for social role implied by an outfit. Participants were randomly assigned to each condition and gender was counterbalanced across the groups.

16.5.2 Results

In the study, 64 people took part. Half of them were women and half of them men. The subjects' age ranged from 18 to 30 years, with an average of approximately 22 years. Subjects claiming prior knowledge of Milgram's experiment in the questionnaire filled in after the experiment were eliminated from the sample and replaced with other participants.

Quantitative Results. The dependent variable examined was the voltage level of the strongest administered shock (further referred to as "resignation level"). The variable has shown a bimodal distribution. The participants either stopped the experiment at ca. 150-180 Volts, or they completed it by administering the 450V shock. Fig. 16.1 shows frequency distribution of level of resignation across the entire sample.

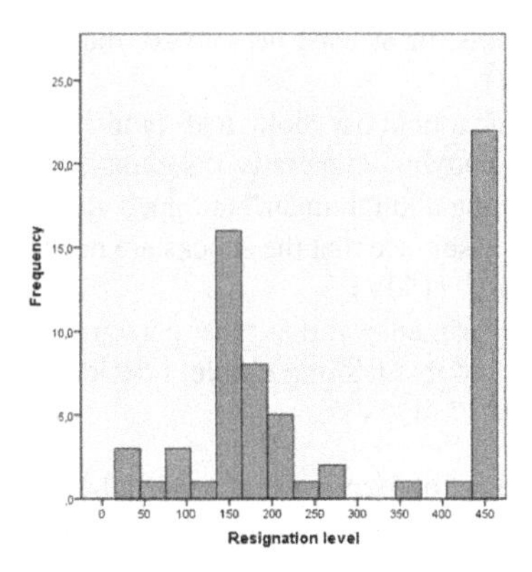

Fig. 16.1 Frequency distribution of the highest used voltage across the entire sample (n=64)

Because of the bimodal distribution two methods of statistical analysis of results were used: a non-parametric U Mann-Whitney's test and additionally ANOVA. U Mann-Whitney's test has shown very strong tendencies in both variables, which were on the border of statistical significance but has not crossed it. It was p=0,056 for name (lower level of resignation for participants controlling avatar with their name than with a number) and p=0,057 for outfit (lower for civilian than military). Additional ANOVA analysis has shown significant main effect of name (p=0,048, F=2,884) and insignificant effect of outfit.

The questionnaire contained a series of questions concerning computer games. None of the answers constituted a significant predictor variable of voltage level at which the participant will decide to quit the experiment. Regardless of how much or how little experience the participants had with computer games (question 5), their game genre preferences (question 6) or their attitude regarding computer entertainment (question 8), none of these factors achieved statistical significance. Similar results were observed in case of demographic variables, i.e. age and education.

Qualitative Analysis. The final questionnaire contained two open questions, in order to give the participants a chance to express their opinions concerning the experiment. The first question was split into two parts. In the first part, the question "Did you administer the strongest shock?" was asked. The second part prompted the participant to explain why. The second question dealt with the subject of how the participants experienced the whole situation. The question was: "What are your feelings concerning this experiment?"

Regardless of whether or not the participant completed the experiment, a large majority of answers referred to the learner as if he was a person (anthropomorphism) and either cited his will, or took the form of excuses. Some example answers supporting this claim:

"The subject had health problems, or at least he said so, there was too much risk." (experiment stopped after 210V)

"I feel this is inhuman. Physical (and psychological) punishment does not improve memory, nor does it help with studying. Generally it's good that it's just a computer, but even in the game it's disgusting and inhuman." (stopped after 210V)

"Despite the experimenter's assurance that the shocks are not dangerous to Christopher, I didn't want him to be hurt." (150V)

"Since I asked the virtual experimenter if that other person can quit the experiment several times, and once I found out that it's impossible, I decided that it would be best for him to get it over with quickly." (450V)

Some answers seemed to compare the experiment to a real-life situation, suggesting that the participants, while playing the game, automatically experienced the simulation on par with a situation in the real world:

"At first I didn't feel any emotions. But once the subject started to complain about the administered punishment, I decided that I can't continue, since I imagined this situation in reality." (180V)

"Because I couldn't cause such a strong pain to another human being, even though I knew that he was not a real person." (180V)

In the group of people who administered the strongest shock, statements showing extreme detachment from computer-generated learner's feelings and humanity were very rare, but present nonetheless. Some examples are given below:

"Because I treated the experiment as unreal, and I wanted to complete it." (450 V)
[Why did you complete the experiment?] "Because it's just a game." (450 V)

Despite the presence of such answers, the dominating pattern was for the participants to refer to the (computer-generated) character's feelings in a more or less apparent manner, usually along with a comment stating that the situation was not real and so there was no reason to feel anything. Nevertheless, the very fact that participants asked themselves about moral consequences of their actions, leads to an assumption that some degree of doubt, despite full awareness of the fact that the situation was artificial, must have arisen.

"I wanted to complete the experiment, and at the same time I knew that the shocks will not hurt the learner." (450 V)
"I wanted to stop it earlier, but I wanted to do the task properly." (450 V)

An interesting case is that of a person who quit the experiment after administering a 270-volt shock. The reason why did he stop was as follows:

"I quit when the NPC started to beg to stop for some time. The voltage level began to rise dramatically."

This answer suggests that the participant fully entered into the spirit of the story and treated it equally with reality. In this context the answer given to the second question "What are your feelings concerning the experiment?" is surprising: "The distance between the real and virtual worlds". Most likely two opposed tendencies arose in him, who tried both to treat the situation as a genuine social event and, at the same time, was fully aware that, objectively, it cannot be classified as such. Contradictory answers were given as a consequence of this clash.

To sum up the analysis of the participants' statements it should be noted that most of them showed very apparent tendencies towards anthropomorphizing the learner. Even those who completed the experiment (those who administered the 450-volt shock) very often did not depersonalize him, but instead they transferred their guilt on the experimenter, or on the fact that the whole situation was not real (but it is worth noting that the presence of sense of guilt implies that someone—the learner in this case—was hurt). Those tendencies to humanize are well summarized by these three quotations:

> [What are your feelings concerning this experiment?] "Fear of the answerer's death." (after administering the 450-volt shock)
> [Why did you quit?] "Because it's inhuman." (210 V)
> "I was causing him pain and I feel bad about it." (270 V)

16.6 Discussion and Conclusions

Two main conclusions can be drawn from the presented results. First, qualitative results strongly support claims of the *Media Equation Theory* [21] that we tend to perceive virtual social situations as real which stands in line with the results of another Milgram-inspired virtual experiment conducted in 2006 by Mel Slater and his group [23]. What can become a surprise is the fact that the described experiment was very far from being realistic. The signs of pain were presented only as a floating text above the learners' figure, yet still were emotionally convincing enough to force approximately two thirds of participants to quit. It stands in strong contrast to what computer games are usually seen as—an area for pointless and inhuman violence. Moreover, previous experience with computer entertainment had no influence on the results. Possibly the explanation of this incoherence could lie in the characteristics of a task. In violent computer games, the main objective is to kill and this objective is more (for example in simulators of war or alien invasion) or less justified (in computer games like Postal or Carmageddon), while in other games it can even become a serious moral dilemma (for example in Planescape: Torment). Similarly to the aforementioned research by Zadro et al. [30] where participants felt excluded by digital confederates, in a specific contexts we spontaneously and naturally get engaged into virtual social situations. In Chap. 3 of this volume, Tarek Besold writes about the differences between the formal logics and logics of the human mind. Presented results show how differently human social mind works in comparison to the objective available data. It does not matter that a learner does not feel anything at all—we still sympathize with him and feel sorry for his (illusory) pain.

Second conclusion. Recent research within the fields of psychology, neuroscience and media studies highlight the problem of relation between the user (gamer) and the avatar being under his or her control. Approaches to this case combine the latest discoveries concerning the body image and self concept with a more social approach.

Overview of the results of the latest experiments suggests that user-controlled avatars can, at least in some cases, become parts of the representations of physical body (there are doubts whether it can be incorporated into human body representation). On the other hand, the degree of such fusion might be mediated by situational (and cultural) cues which are processed by higher cognition. This research was aimed to test whether such cues can cause differences in perceived relation to the avatar and then succeeding differences in behavior. The results are not concluding but show strong but insignificant influence in the predicted direction. It means that manipulating the degree of unification with an avatar is presumably possible but it requires further investigation especially within the neuroscientific domain in order to support or discard this insight.

Final conclusion. In regard to artificial intelligence in general it seems that people tend to naturally care about what artificial agents "feel" even though these agents actually cannot feel anything at all. Taking into account all the possible risks associated with developing artificial intelligence (which have been discussed in the Beyond AI 2011 keynote speech by Anders Sandberg), maybe it is a good time to start thinking about implementing an empathetic social mind into it. If we do care what an artificial being feels then maybe it should also begin to care about us.

References

1. Altabe, M., Thompson, J.K.: Body Image: A Cognitive Self-Schema Construct? Cognitive Therapy and Research 20(2), 171–193 (1996)
2. Bailenson, J.N., Blascovich, J.J.: Avatars. In: Bainbridge, W.S. (ed.) Berkshire Encyclopedia of Human-Computer Interaction, pp. 64–68. Berkshire Publishing Group (2004)
3. Bem, D.J.: Self-Perception Theory. In: Berkovitz, L. (ed.) Advances in Experimental Social Psychology, vol. 6, pp. 1–62. Academic Press, New York (1972)
4. Bickmore, T., Cassel, J.: Social Dialogue with Embodied Conversational Agents. In: van Kuppevelt, J., Dybkjaer, L., Bernsen, N. (eds.) Natural, Intelligent and Effective Interaction with Multimodal Dialogue Systems, pp. 23–54. Kluwer Academic, New York (2004)
5. Blanke, O., Mohr, C., Michel, C.M., Pascual-Leone, A., Brugger, P., Seeck, M., Landis, T., Thut, G.: Linking Out-of-Body Experience and Self Processing to Mental Own-Body Imagery at the Temporoparietal Junction. Journal of Neuroscience 25(3), 550–557 (2005)
6. Carruthers, G.: Types of Body Representation and the Sense of Embodiment. Consciousness and Cognition 17(4), 1302–1316 (2008)
7. de Preester, H., Tsakiris, M.: Body-Extension versus Body-Incorporation: Is There a Need for a Body-Model? Phenomenology and the Cognitive Sciences 8(3), 307–319 (2009)
8. de Vignemont, F.: Body Schema and Body Image-Pros and Cons. Neuropsychologia 48(3), 669–680 (2010)
9. de Vignemont, F.: Embodiment, Ownership and Disownership. Consciousness and Cognition 20(1), 82–93 (2011)
10. Gallagher, S.: Philosophical Conceptions of the Self: Implications for Cognitive Science. Trends in Cognitive Sciences 4(1), 14–21 (2000)
11. Gallagher, S.: How the Body Shapes the Mind. Oxford University Press (2005)

12. Gazzola, V., Rizzolatti, G., Wicker, B., Keysers, C.: The Anthropomorphic Brain: The Mirror Neuron System Responds to Human and Robotic Actions. Neuroimage 35(4), 1674–1684 (2007)
13. Ganesh, S., van Schie, H.T., de Lange, F.P., Thompson, E., Wigboldus, D.H.J.: How the Human Brain Goes Virtual: Distinct Cortical Regions of the Person-Processing Network Are Involved in Self-Identification with Virtual Agents. Cerebral Cortex (2011), doi:10.1093/cercor/bhr227
14. Leenhardt, M.: Do Kamo. La personne et le mythe dans le monde melanesien. Gallimard (1947)
15. Maravita, A., Iriki, A.: Tools for the Body (Schema). Trends in Cognitive Sciences 8(2), 79–86 (2004)
16. Marchetti, C., Della Sala, S.: Disentangling the Alien and Anarchic Hand. Cognitive Neuropsychiatry 3(3), 191–207 (1998)
17. Markus, H.: Self-Schemata and Processing Information about the Self. Journal of Personality and Social Psychology 35(2), 63–78 (1977)
18. Melzack, R.: Phantom Limbs and the Concept of a Neuromatrix. Trends in Neurosciences 13(3), 88–92 (1990)
19. Milgram, S.: Obedience to Authority: An Experimental View. Harper & Row (1974)
20. Morewedge, C.K., Preston, J., Wegner, D.M.: Timescale Bias in the Attribution of Mind. Journal of Personality and Social Psychology 93(1), 1–11 (2007)
21. Reeves, B., Nass, C.: The Media Equation: How People Treat Computers, Television, and New Media like Real People and Places. Cambridge University Press (1996)
22. Schwoebel, J., Branch Coslett, H.: Evidence for Multiple, Distinct Representations of the Human Body. Journal of Cognitive Neuroscience 17(4), 543–553 (2005)
23. Slater, M., Antley, A., Davidson, A., Swapp, D., Guger, C., Barker, C., Pistrang, N., Sanchez-Vives, M.V.: A Virtual Reprise of the Stanley Milgram Obedience Experiments. PLoS ONE 1(1) (2006), doi:10.1371/journal.pone.0000039
24. Synofzik, M., Vosgerau, G., Newen, A.: Beyond the Comparator Model: a Multifactorial Two-Step Account of Agency. Consciousness and Cognition 17(1), 219–239 (2007)
25. Synofzik, M., Vosgerau, G., Newen, A.: I Move, Therefore I Am: A New Theoretical Framework to Investigate Agency and Ownership. Consciousness and Cognition 17(2), 411–424 (2008)
26. Tsakiris, M., Costantini, M., Haggard, P.: The Role of the Right Temporo-Parietal Junction in Maintaining a Coherent Sense of One's Body. Neuropsychologia 46(12), 3014–3018 (2008)
27. Turing, A.: Computing Machinery and Intelligence. Mind 49(236), 433–460 (1950)
28. Waytz, A., Epley, N., Cacioppo, J.T.: Social Cognition Unbound: Insights Into Anthropomorphism and Dehumanization. Current Directions in Psychological Science 19(1), 58–62 (2010)
29. Yee, N., Bailenson, J.N.: The Proteus Effect: The Effect of Transformed Self-Representation on Behavior. Human Communication Research 33(3), 271–290 (2007)
30. Zadro, L., Williams, K.D., Richardson, R.: How Low Can You Go? Ostracism by a Computer is Sufficient to Lower Self-Reported Levels of Belonging, Control, Self-esteem and Meaningful Existence. Journal of Experimental Social Psychology 40(4), 560–567 (2004)
31. Alicebot, `http://alicebot.blogspot.com/`
32. GRETA, `http://perso.telecom-paristech.fr/~pelachau/Greta/`
33. Jabberwacky, `http://www.jabberwacky.com/`

Index

I

K

L

M

N

P

R

S

T

U

V

W